IMAGES OF ASIA

The Birds of Sulawesi

Titles in the series

At the South-East Asian Table
ALICE YEN HO

Balinese Paintings
A. A. M. DJELANTIK

Bamboo and Rattan
JACQUELINE M. PIPER

Betel Chewing Traditions
in South-East Asia
DAWN F. ROONEY

The Birds of Java and Bali
DEREK HOLMES and STEPHEN NASH

The Birds of Singapore
CLIVE BRIFFETT and SUTARI BIN SUPARI

The Birds of Sulawesi
DEREK HOLMES and KAREN PHILLIPPS

Borobudur
JACQUES DUMARÇAY

Burmese Dance and Theatre
NOEL F. SINGER

Burmese Puppets
NOEL F. SINGER

Folk Pottery in South-East Asia
DAWN F. ROONEY

A Garden of Eden: Plant Life in
South-East Asia
WENDY VEEVERS-CARTER

Gardens and Parks of Singapore
VÉRONIQUE SANSON

The House in South-East Asia
JACQUES DUMARÇAY

Images of the Buddha in Thailand
DOROTHY H. FICKLE

Imperial Belvederes: The Hill Stations
of Malaya
S. ROBERT AIKEN

Indonesian Batik
SYLVIA FRASER-LU

Javanese Gamelan
JENNIFER LINDSAY

Javanese Shadow Puppets
WARD KEELER

The Kris
EDWARD FREY

Lao Textiles and Traditions
MARY F. CONNORS

Life in the Javanese Kraton
AART VAN BEEK

Mammals of South-East Asia
EARL OF CRANBROOK

Musical Instruments of South-East Asia
ERIC TAYLOR

Old Bangkok
MICHAEL SMITHIES

Old Kuala Lumpur
J. M. GULLICK

Old Luang Prabang
BETTY GOSLING

Old Malacca
SARNIA HAYES HOYT

Old Manila
RAMÓN MA. ZARAGOZA

Old Penang
SARNIA HAYES HOYT

Old Singapore
MAYA JAYAPAL

Rice in South-East Asia
JACQUELINE M. PIPER

Sarawak Crafts
HEIDI MUNAN

Songbirds in Singapore
LESLEY LAYTON

Traditional Festivals in Thailand
RUTH GERSON

The Maleo, p. 15.

The Birds of Sulawesi

DEREK HOLMES

Illustrated by
KAREN PHILLIPPS

KUALA LUMPUR
OXFORD UNIVERSITY PRESS
OXFORD SINGAPORE NEW YORK
1996

Oxford University Press

Oxford New York
Athens Auckland Bangkok Bombay
Calcutta Cape Town Dar es Salaam Delhi
Florence Hong Kong Istanbul Karachi
Madras Madrid Melbourne Mexico City
Nairobi Paris Shah Alam Singapore
Taipei Tokyo Toronto

and associated companies in
Berlin Ibadan

Oxford is a trade mark of Oxford University Press

Published in the United States
by Oxford University Press, New York

© Oxford University Press 1996
First published 1996

All rights reserved. No part of this publication may be reproduced,
stored in a retrieval system, or transmitted, in any form or by any means,
without the prior permission in writing of Oxford University Press.
Within Malaysia, exceptions are allowed in respect of any fair dealing for the
purpose of research or private study, or criticism or review, as permitted
under the Copyright Act currently in force. Enquiries concerning
reproduction outside these terms and in other countries should be
sent to Oxford University Press at the address below

ISBN 983 56 0005 8

British Library Cataloguing in Publication Data
Data available

Library of Congress Cataloging-in-Publication Data
Data available

Typeset by Expo Holdings Sdn. Bhd., Malaysia
Printed by Kyodo Printing Co. (S) Pte. Ltd., Singapore
Published by the South-East Asian Publishing Unit,
a division of Penerbit Fajar Bakti Sdn. Bhd.,
under licence from Oxford University Press,
4 Jalan U1/15, Seksyen U1, 40000 Shah Alam,
Selangor Darul Ehsan, Malaysia

Acknowledgements

THE author wishes to thank Chris Escott and Dick Watling, both of whom accompanied him in the field during early trips to Sulawesi and introduced him to some of the island's endemic birds. Both Chris and Dick kindly read through an early draft of this manuscript and their comments were most helpful, although responsibility for the accuracy of the text rests with the author. David Bishop, co-author of the forthcoming *A Guide to the Birds of Wallacea: Sulawesi, Moluccas and Lesser Sundas* (Dove Publications), has also generously provided assistance with reading the text and providing additional data on several species. Use was also made of published taped recordings of Sulawesi bird songs made by David Gibbs and Steve Smith. Sulawesi is now a prime venue for bird-watchers, most of whom submit their observations to the Indonesian Ornithological Society, and the author has drawn freely from their published records.

Karen Phillipps thanks Dr James Lazell of the Conservation Agency for setting up the 'Sulawesi Expedition'. Dr James Dolan and David Rimlinger of the San Diego Zoological Society and in Hong Kong Dr Ken Searle and Dr John MacKinnon provided access to their notes, libraries, and live bird collections, without which the illustrations would have been much less accurate.

Jakarta   DEREK HOLMES
October 1995   KAREN PHILLIPPS

Contents

Acknowledgements *v*
Introduction *ix*

Grebes *1*
Frigatebirds *1*
Cormorants *2*
Boobies *3*
Herons, Egrets, and
   Bitterns *3*
Storks and Ibises *7*
Bird of Prey *8*
Ducks *13*
Megapodes *14*
Crakes and Rails *16*
Jacanas *19*
Shorebirds *19*
Stilts *21*
Thick-knees *21*
Terns *22*
Pigeons and Doves *23*
Parrots *29*
Cuckoos *32*
Owls *35*
Nightjars *37*
Swifts *38*
Kingfishers *40*
Bee-eaters *44*
Rollers *45*
Hornbills *46*

Woodpeckers *47*
Pittas *47*
Swallows *48*
Wagtails and Pipits *49*
Cuckoo-shrikes and Trillers *49*
Bulbuls *52*
Thrushes *52*
Babblers *54*
Malia and Geomalia *55*
Warblers *55*
Flycatchers *57*
Whistlers *61*
Flowerpeckers *62*
Sunbirds *62*
White-eyes *64*
Honeyeaters *65*
Sparrows, Munias, and
   Finches *66*
Starlings and Mynas *68*
Orioles *70*
Drongos *71*
Wood-swallows *71*
Crows *72*

Appendix *73*
Further Reading *82*
Index to Genera, Systematic
   Section *83*

Introduction

For the bird-watcher, Sulawesi is unequalled. The bird list of the Sulawesi region comprises 380 species, perhaps not a very high total, but no less than 96 of these are endemic to the region, an incredible 25 per cent of the avifauna, while 115 are endemic within Indonesia. Ornithologists from all over the world place the island high on their list of priorities. Yet, until very recently, there has never been a book on the island's birds, or a readily available source from which to identify the endemics.

The present volume does not attempt to be comprehensive; it is designed as a popular work to introduce the bird-watcher, and all those with a lively interest in their natural surroundings, to the island's birds, and to most of the endemics that can be readily encountered. It is particularly designed to provide the means for Indonesians to take an interest in their unique natural environment, as a major step in preserving this remarkable heritage for future generations.

The Sulawesi region, for the purposes of this book, refers to the avifaunal region, rather than to administrative boundaries. The region covers the main island of Sulawesi and the Sangihe, Talaud, Peleng, and Banggai island groups, as well as the nearby Sula Islands which are part of Maluku province. The Sula Islands have a fauna that is transitional between Sulawesi and Maluku, but the affinities are more closely aligned to Sulawesi.

The unique fauna results from Sulawesi's position and evolution, in the complex geological zone that connects the Sundaic and Sahulian (Australo-Papuan) subcontinents. To biologists, this zone of transitional faunas and endemism is known as Wallacea, named after the biologist-explorer Alfred Wallace. Wallacea broadly includes Sulawesi, Maluku, and Nusa Tenggara (the latter often known as the Lesser Sundas).

The term 'Sundaic' here refers to the Sundanese biogeographical region, a wide continental land that was exposed during geological periods of lowered sea-levels, comprising western Indonesia (the

INTRODUCTION

islands of Java, Sumatra, and Borneo), and the Malay Peninsula. This term is used in order to avoid confusion with various other meanings of the word 'Sundanese'.

The boundary between the Sundaic and Wallacean faunas is well defined. It runs between Kalimantan and Sulawesi, and between Bali and Lombok. Likewise the boundary between the Wallacean and Papuan faunas lies between Maluku and Irian Jaya (though the Aru Islands lie on the Sahulian shelf and have a Papuan fauna).

The differences are readily apparent. An ornithologist who studies birds in the Kutai National Park in East Kalimantan, and then flies across to Palu in Sulawesi and ascends into Lore Lindu National Park, might believe he has flown to a different continent. In the former, he will have been observing or listening to a wide variety of hornbills, barbets, woodpeckers, and babblers, with just a few pigeons and parrots. In the latter, he will face a puzzling bewilderment of pigeons and parrots, and very few species of hornbills, woodpeckers, and babblers (all endemic); the incessant calls of the barbets will be absent. The number of species in the sample groups listed below illustrates the differences.

	Kalimantan	*Sulawesi*
Green Pigeons	5	2
Fruit-doves	1	4
Imperial Pigeons	3	6
Other pigeons	6	11
Parrots	4	15
Hornbills	8	2
Barbets	8	0
Woodpeckers	17	2
Broadbills	6	0
Bulbuls	22	1 (+2 introduced)
Babblers	35	1
Flowerpeckers	11	3
Sunbirds, Spiderhunters	16	5
Honeyeaters	0	3
Starlings, Mynas	3	11

INTRODUCTION

These figures are not definitive, and feral species and some small island species are excluded. They may indeed slightly mask the real variations, because so few of the species are common to both regions. The impression is also given that Sulawesi has many fewer species of birds. This is true, but the majority of bird species found in Kalimantan also occur in Java, Sumatra, and Malaysia, whereas the Sulawesi birds often have much more restricted ranges.

Most of the author's experience has been in the Sundaic region. To him, the bird sounds of the Sundaic forests seem homely and familiar, and generally cheerful. In contrast, an overriding impression of the Sulawesi forests is one of melancholy. This may be rather intangible, but the cooing of the pigeons, the rolling notes of the Black-billed Koel, the hollow notes of the Bay Coucal, and the slurred whistles of the Red-bellied Pitta, all carry a distinctly mournful effect. Add to this combination the raucous screeching of a flock of parrots flying overhead, and perhaps the deafening trumpet calls of the Channel-billed Cuckoo or the braying of the Maleo, and the Sundaic bird-watcher indeed feels on alien ground. It is an exciting place to study birds!

No specific localities are recommended, as Sulawesi or Wallacean endemics are found everywhere. Obviously the nature reserves and national parks such as Tangkoko–Batuangas, Dumoga–Bone, Lore Lindu, and Morowali will be the main destinations, but some of these are quite remote. Good bird-watching can be experienced anywhere in the region, including the more lightly wooded areas of the south. The avifauna is still rather poorly known, and data are required from all corners of the island, as well as the offshore islands. The eastern and south-eastern peninsulas are particularly poorly studied, but even the ornithologist based in Ujung Pandang can make valuable contributions. The opportunities to study birds on Sulawesi are endless, and all such studies may be valuable.

The book illustrates 142 species in colour, and a further 22 species in black-and-white, accompanied by brief descriptions in the text. Representative examples are selected from the majority of the bird families, particularly the more distinctive and colourful members and the endemic species. These descriptions cover 161 species, but reference is made to a further 88 in order to illustrate the diversity of forms that the more serious ornithologist will need to study.

The scientific and English names are all derived from the Indonesian checklist of birds.* Alternative names are given where appropriate. As the nomenclature is confusing, it is useful to become familiar with the scientific names. However off-putting these may seem, they change less frequently than the English names, and in the absence of definitive Indonesian names for most birds, they provide a valuable international language for ornithologists. It is to be hoped that with the recent publication of this checklist, English names will now become standardized, and that the compilation of Indonesian names will soon be complete.

The length measurements given in the text are provided only as a general guide. There are considerable inconsistencies between different authors and field guides, which probably reflect the actual variations that occur in the wild, according to regional variations, stage of moult, etc. The measurements are mainly useful as a qualitative comparison between species. The Appendix provides a list of the resident birds in the Sulawesi region, tabulated for island groups, and it annotates those that are endemic. A full list of the avifauna, including migrant species, can be found in the Indonesian bird checklist.

*The Birds of Indonesia: A Checklist (Peters' Sequence), by Paul Andrew. Kukila Checklist No. 1, published by the Indonesian Ornithological Society, Jakarta, June 1992.

Note

Abbreviations used on Plates:

♂	Male
♀	Female
Juv.	Juvenile
Br.	Breeding plumage
Non-br.	Non-breeding plumage

Grebes

RED-THROATED LITTLE GREBE
Tachybaptus ruficollis (25 cm)   Plate 3a

Little Grebes are small, squat, duck-like birds with short, pointed bills, often seen swimming and diving in the open waters of lakes and swamps. Dark grey-brown above and whitish below, they develop a chestnut throat and foreneck in breeding plumage. The call often reveals their presence when hidden in the water-side vegetation, a sharp trill that lasts up to three seconds, slightly rising and falling. This and a closely related form occur widely through Indonesia, and from Africa and Europe to Australasia. It is a common resident in the Tempe lake system of South Sulawesi.

Frigatebirds

THE Banda Sea of Maluku holds several colonies of sea birds on small, remote islands, including tropic birds, boobies, and frigatebirds, but in general, sea birds are rather scarce around Sulawesi waters. Most often seen are frigatebirds, and various terns (see p. 22).

LESSER FRIGATEBIRD
Fregata ariel (80 cm)

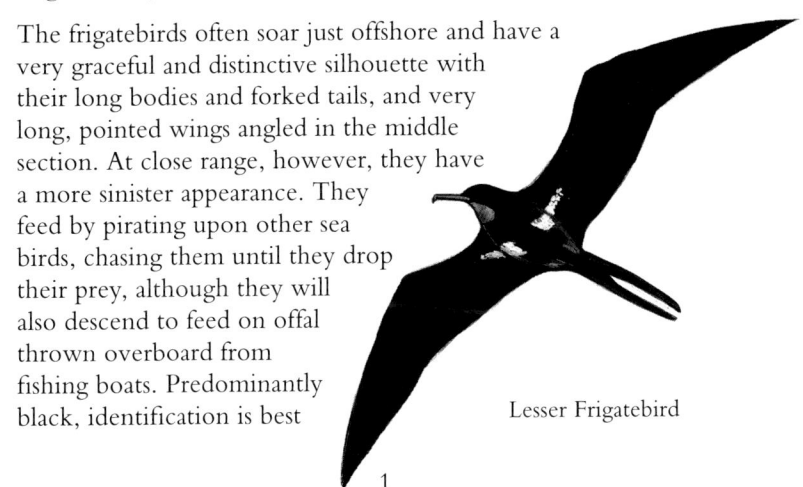

The frigatebirds often soar just offshore and have a very graceful and distinctive silhouette with their long bodies and forked tails, and very long, pointed wings angled in the middle section. At close range, however, they have a more sinister appearance. They feed by pirating upon other sea birds, chasing them until they drop their prey, although they will also descend to feed on offal thrown overboard from fishing boats. Predominantly black, identification is best

Lesser Frigatebird

left to the experts, as there is a varying distribution of white in the plumage according to species, sex, and age of the bird. The males have a red throat pouch which can be inflated during display, although this is not readily seen in flight.

This species and the GREAT FRIGATEBIRD *F. minor* (90 cm) are those most often encountered in Sulawesi. Both the adult males are black, but the Lesser has distinctive white patches beneath the base of the wings. Only the Great Frigatebird breeds regularly in Indonesian waters, although the Lesser Frigatebird has bred on Gunung Api.

Cormorants

CORMORANTS are black or black-and-white, long-billed, long-necked, wedge-tailed water-birds that feed on fish by diving from the surface. They roost gregariously on trees, often with their wings held out to dry in a witch-like stance. Both the cormorants found in Sulawesi are Australasian species, near the northern limit of their range.

LITTLE PIED CORMORANT
Phalacrocorax melanoleucos (60 cm)

The Little Pied Cormorant is common on inland lakes and marshes, particularly on the Tempe lake system, and is at once identified by its cleanly divided black upper-parts and white under-parts. It should not be confused with the Pied Heron (see p. 5).

Two other species are common locally, the LITTLE BLACK CORMORANT *P. sulcirostris*, which is the same size but all black, and the ORIENTAL DARTER *Anhinga melanogaster* (90 cm). The latter bird has an extremely long, sinuous neck, from which it is sometimes known as 'Snakebird'; often only its neck is seen above the surface of the water.

Little Pied Cormorant

Boobies

BROWN BOOBY
Sula leucogaster (75 cm)

The Brown Booby is a huge, cigar-shaped bird with long, narrow wings and long, heavy bill, that is related to the gannets of the temperate seas. Mainly brown, the adult has a white belly sharply demarcated from the brown breast. Boobies have a direct flight, often low over the surface of the water, with steady wing beats alternating with short glides, and they feed by shallow dives on half-closed wings, often from a low angle (gannets, in contrast, are mainly white, and make vertical dives). There are three species in Indonesian waters, but the Brown Booby is the one most often encountered close to shore.

Brown Booby

Herons, Egrets, and Bitterns

THESE are tall wading birds with long necks and legs, that feed by stalking through shallow waters or mud, stabbing at small fish or crustaceans with their dagger-like bills. They fly with a steady beat of rounded wings, with the neck curved back in a 'S'-shape. The egrets are predominantly white, and often feed in rice-fields, while the bitterns have cryptic colours and are secretive birds of dense vegetation. The status of many of these birds in Sulawesi is complex, with both resident and migrant forms.

The common heron on fresh-water lakes and swamps is the PURPLE HERON *Ardea purpurea* (97 cm), which is grey and purple-brown in colour. It can be seen around the Tempe lakes as well as on smaller wetland areas anywhere in the region. A very large and solitary grey heron of coastal areas is the GREAT-BILLED HERON *A. sumatrana* (115 cm).

CATTLE EGRET
Bubulcus ibis (50 cm)

The Cattle Egret is very common in wet meadows and rice-fields, and is partial to feeding around the feet of buffaloes, even at times perching on their backs. It has a yellow bill and, in breeding plumage, orange-buff plumes on the fore-parts.

Also very common in the rice-fields is the JAVAN POND-HERON *Ardeola speciosa* (45 cm). This is a more squat bird with rather cryptic brown and white colours, but with startling white wings when it takes flight. In breeding plumage, the back is black and the neck and breast are cinnamon. Both this and the Cattle Egret breed on Sulawesi, although numbers may be augmented by seasonal visitors.

Cattle Egret

The white egrets range in size from the tall GREAT EGRET *Casmerodius albus* (90 cm), and the slightly smaller INTERMEDIATE EGRET *Egretta intermedia* (70 cm), to the graceful and delicate LITTLE EGRET *E. garzetta* (60 cm). These occur around estuaries, mud-flats, and the larger wetlands, but among these three, only the Little Egret is known to breed on Sulawesi.

PIED HERON
Egretta picata (45 cm)   Plate 1a

The Pied Heron is locally common on fish-ponds and fresh-water swamps, more particularly in South Sulawesi, where it is resident. It is an attractive bird, slate grey, with white neck and breast plumes, and a black cap to the head; there is a short crest, and the bill and legs are yellow. Young birds have the head all white. Its range extends from Sulawesi to New Guinea and North Australia.

STRIATED HERON
Butorides striatus (47 cm)   Plate 1b

This is an inconspicuous small, streaky grey heron with a black cap and crest. It is a solitary, furtive bird that feeds quietly with a hunched stance in mangroves or waterside vegetation, occasionally along streams far into the forest. It sometimes roams out on to open beaches and reefs, but will quickly take flight if approached, uttering a long 'skeoh' if disturbed suddenly.

RUFOUS NIGHT-HERON
Nycticorax caledonicus (60 cm)   Plate 1d

Night-herons are rather squat herons with heavy heads and moderately thick bills, that usually roost by day, and fly out to feed at night in the rice-fields, with harsh 'kwak' calls. Plate 1d illustrates the distinctive plumage of the Rufous Night-heron. Note the two white streamers from the crown of the breeding adult. Wallacea forms the boundary between the Rufous Night-heron of Australasia and eastern Indonesia and the much more widely distributed BLACK-CROWNED NIGHT-HERON *N. nycticorax*, which occurs in western Indonesia. The latter bird, which is illustrated in Plate 1c, is blue-black above and white below. It is uncommon in Sulawesi, but it breeds in the same colonies and occasionally the two species hybridize. The young of both species are streaked brown and difficult to tell apart.

CINNAMON BITTERN
Ixobrychus cinnamomeus (40 cm) — Plate 1e

Cinnamon Bitterns live quietly in rice-fields, reed-beds, and dense swamp vegetation, and are most often seen when they take short flights from one feeding area to another. In flight, the drawn-in neck and rounded wings of uniform chestnut colour are distinctive. When perched, the under-parts are seen to be buffy with a dark line down the centre of the breast. This species is a common resident throughout Sulawesi.

YELLOW BITTERN
Ixobrychus sinensis (37 cm) — Plate 1f

The Yellow Bittern is a common migrant to reed-beds and the rice plains during the northern winter. It is distinguished by its paler buffy plumage, with black cap and tail, and the black tips of the wings which are clearly visible in flight.

BLACK BITTERN
Ixobrychus flavicollis (60 cm) — Plate 1g

A larger species of uncertain status, this bittern is blackish in colour with a streaky breast, and a clearly visible buffy-yellow stripe on the sides of the neck. Its habitat includes the margins of wetland forest, mangroves, and the more densely vegetated swamps, as well as the rice plains. Although more common during the northern winter, it seems to be present most of the year, and there may be both northern migrants and either a resident population or southern migrants involved.

Storks and Ibises

STORKS are tall, gaunt birds that differ from herons in having a stouter build and more massive bills, and in their habit of flying and soaring with neck and legs outstretched.

WOOLLY-NECKED STORK
Ciconia episcopus (90 cm)

The Woolly-necked Stork occurs throughout Sulawesi in open wetlands, sometimes in loose parties of a dozen or more. It is black with a white neck and belly, separated by a black breast band; the bill and legs are reddish.

The endangered MILKY STORK *Mycteria cinerea* (100 cm) has been discovered in Sulawesi only quite recently and may also be resident. Most reports are from South Sulawesi, but it has also been seen in the South-east and North. It is a graceful stork of coastal wetlands, with a slightly down-curved bill and whitish plumage, except for a dark breast band and black tail and flight feathers.

Woolly-necked Stork

GLOSSY IBIS
Plegadis falcinellus (65 cm)

The Glossy Ibis is a smaller bird, and its long, down-curved bill contributes to the total length quoted above. It is a glossy dark chestnut,

Glossy Ibis

appearing black in a poor light. It is a gregarious bird of open wetlands, including the more swampy rice-fields, especially in the south. It is believed to be resident, but breeding has not yet been confirmed. It moves around locally according to water-levels in the swamps. Up to 5,000 birds have been counted on the Tempe lakes at certain seasons. In Central Sulawesi, the author found it common in the Besoa valley.

Birds of Prey

THE birds of prey, or raptors, are hooked-billed predators that seek their prey from the air by a variety of methods, usually dropping on their prey feet first. The methods range from soaring (kites, buzzards, eagles), quartering low over the ground (harriers), swift flight through the tree-tops (sparrow-hawks), and steep dives (falcons), while a few species hover in open country. There are twenty-three resident species on Sulawesi, of which no less than six are endemic, but the identification of many of these is best left to the expert, particularly the four endemic sparrow-hawks in the genus *Accipiter*.

OSPREY
Pandion haliaetus (55 cm)   Plate 2a

Sulawesi has both a resident and a rare migrant form of this worldwide species. It is a fish-hawk, that feeds by quartering back and forth over lakes and coastal waters, briefly hovering and then plunging feet first to catch its prey. Mainly white on the head and underparts and brown above, it has a broad black band through the eye, a black patch on the bend of the wing, rather long and slender wings, and a square tail. It is most often seen along wooded coasts.

WHITE-BELLIED SEA-EAGLE
Haliaeetus leucogaster (70 cm)

Sulawesi has three other fish-eating raptors, but only the White-bellied Sea-eagle is common. Superficially, this may resemble the Osprey, but it is larger, has broader wings, and a wedge-shaped tail. The flight feathers and base of the tail are black, but it has no black band through the eye. Primarily a coastal bird, seen most often soaring just offshore, or resting in a commanding tree near the beach, it also occurs and breeds around inland lakes.

White-bellied Sea-eagle

BLACK KITE
Milvus migrans (65 cm)

The Black Kite is a common scavenger through much of Asia and Australia, but it is a rather enigmatic bird in Indonesia, being generally only a very rare migrant. In Sulawesi, however, there is a local resident population, although perhaps subject to local movements. It can be seen soaring effortlessly, singly or in loose parties, over open country, coasts, and fish-ponds, on rather long wings, with a diagnostic forked tail. The tail is constantly adjusted from side to side as the bird rides the currents, and the fork is not seen when the tail is fanned. It is dark brown, sometimes with a paler head and a pale patch in the wings. Found throughout Sulawesi, it is more common in the south, though numbers may be decreasing.

Black Kite

BRAHMINY KITE
Haliastur indus (45 cm)

This is a familiar scavenger of coasts and harbours, and often also inland lakes. The adult is readily identified by its bright chestnut plumage with a shining white but slightly streaky head, neck, and breast. The tail is not forked. Young birds are dull brown and spotted, and can be confusing.

The BLACK-WINGED KITE *Elanus caeruleus* (33 cm) is a small, graceful raptor which is often seen hovering over open country. It is pale grey, whitish below, with a distinctive black patch on the bend of the wing. This and the Spotted Kestrel are the two raptors that normally hunt by hovering.

Brahminy Kite

SULAWESI SERPENT-EAGLE
Spilornis rufipectus (50 cm)   Plate 2b

The Sulawesi Serpent-eagle, which is endemic to the Sulawesi region including the Sula Islands, is closely related to the Crested Serpent-eagle *S. cheela* of the Sundaic region. It is readily identified from the other broad-winged buzzards and eagles that soar over forested terrain by its diagnostic broad white band in the tail and the white band close to the hind margin of the under-wing. When perched, it has a medium-sized crest. The shrill, di- or tri-syllabic call often draws attention to a pair of soaring birds, but variants of this call have been described.

BARRED HONEY-BUZZARD
Pernis celebensis (52 cm)   Plate 2g

The other soaring forest raptors are more difficult to identify, and include the Barred Honey-buzzard, endemic to Sulawesi and the Philippines, and the SULAWESI HAWK-EAGLE *Spizaetus lanceolatus* (60 cm), endemic to Sulawesi and the Sula Islands, which is shown in Plate 2h. The plumage of these two birds is remarkably similar, and both have pronounced banding in the tail. Experts will recognize them primarily by their silhouette, the larger hawk-eagle having strikingly broad wings, while the honey-buzzard has a rather small, pigeon-like head.

The BLACK EAGLE *Ictinaetus malayensis* is a more widely distributed bird that is common in the hills and mountains of Sulawesi, identified by its larger size (70 cm), and black plumage with yellow feet.

A smaller hawk of lightly wooded, hilly country in South Sulawesi is the RUFOUS-WINGED BUZZARD *Butastur liventer* (40 cm), which is dark rufous on the upper surface of the wings and tail. The under-parts are pale, grey on the breast and white on the belly. It flies with shallow wing beats and short glides. Java and South Sulawesi are isolated outposts of this Asian species, but it is not common.

SPOT-TAILED GOSHAWK
Accipiter trinotatus (30 cm)   Plate 2c

The sparrow-hawks and goshawks have short, broad wings and moderately long tails, and hunt by darting flight through the canopy. Outside the forest, they have alternate flapping and gliding flight, and they sometimes also soar. Sulawesi has four resident species, all of them endemic, although the VINOUS-BREASTED SPARROW-HAWK *A. rhodogaster* (see Plate 2d) also occurs on the Sula Islands. These four hawks are so alike that the experts have great difficulty in identifying them, especially when seen darting through the forest canopy. Indeed, they have been described as the greatest identification problem in Sulawesi.

The Spot-tailed Goshawk appears to be the commonest, and can be identified by its diagnostic white spots on the upper-side of the dark tail; these can be seen quite readily. It is bluish-grey above, with a rufous breast and belly. The call consists of about five sharp notes, uttered rather slowly down the scale.

SPOTTED HARRIER
Circus assimilis (55 cm)   Plate 2e

Harriers are readily identified by their shape (medium size with long body, wings, and tail) and habits (gracefully quartering low over open ground with alternating wing beats and glides, with the wings up-turned and the primary feathers splayed when gliding). They are mainly Palaearctic in distribution (Eurasia south to Africa and the Orient), but the Spotted Harrier is an Australian species that has an outpost in Sulawesi. It is a striking bird, light grey above and variably rufous below, finely spotted with white, with dark-tipped wings, fine black bars on the tail, and yellow legs. It is moderately common in Sulawesi, but locally distributed, mainly in grasslands.

SPOTTED KESTREL
Falco moluccensis (32 cm)   Plate 2f

The Spotted Kestrel is another raptor of open or lightly wooded country. It is distinguished by its small size and rather long, narrow wings and tail, but the most characteristic feature of kestrels is their habit of hovering for prolonged periods at one spot, before dropping on their prey (the Black-winged Kite also hovers, see p. 10). The plumage is mainly rich rufous, spotted black, and the bluish-grey tail has a black band near the white tip. The call is a sharp, rapid 'kee-kee-kee-kee-kee'.

Ducks

WANDERING WHISTLING-DUCK
Dendrocygna arcuata (50 cm) Plate 3b

Whistling-ducks, also known as tree-ducks, are noted for their incessant whistled calls. Various species are resident through Indonesia, and this is the most common and widespread form in Sulawesi. It occurs sometimes in quite large numbers in all open swampland and the wetter rice-fields. Mainly dark brown to blackish, in good light they are seen to have a dark cap, pale face, chestnut on the wings, and quite striking white flanks and sides to the rump. The whistled call can be rendered as variants of 'tititipi'. It occurs from Sulawesi and East Java to Australia.

SPOTTED WHISTLING-DUCK
Dendrocygna guttata (43 cm) Plate 3c

The range of this smaller whistling-duck extends from the Philippines to beyond New Guinea, but it also occurs on Sulawesi, where it is perhaps confined to the northern and eastern arms of the island. However, there are few recent records. The heavy white spotting especially on the flanks is diagnostic, but it lacks the white on the rump.

SUNDA TEAL
Anas gibberifrons (40 cm) Plate 3d

The other common resident duck, though in smaller numbers, is the Sunda Teal. It is a small duck with rather indistinct, mottled greyish-brown plumage, paler on the head and under-parts. In flight, there is white at the base of the under-wing, and a white, black, and green patch (the speculum) in the upper-wing. It is found on a variety of waters, from rather swift rivers on the forest margins to the swamps and lakes of the lowlands and coasts, where it feeds by dabbling and 'up-ending'. Formerly considered a race of the Grey Teal, taxonomists currently separate the Grey Teal *A. gracilis* of Australia as a different species. However, the latter duck also migrates to eastern Indonesia, and the relationships of the two species in our region have yet to be determined.

PACIFIC BLACK DUCK
Anas superciliosa (50 cm)   Plate 3e

The Pacific Black Duck is a more local resident, found on the Tempe lakes and coastal areas of the south. The author has also encountered it in the Besoa valley in Lore Lindu National Park. It has mottled dark brown plumage, with well-marked stripes on the face, a green patch on the upper-wing and much white beneath the wing.

GARGANEY
Anas querquedula (40 cm)   Plate 3f

While Indonesia does not have the big flocks of northern duck that migrate to Africa and India during the northern winter, recent surveys have shown that the shallow lakes of South Sulawesi are an important wintering ground for the Garganey, with over 10,000 recorded on Lake Tempe during November. The male of this small duck has a pronounced long, white eyebrow, and its wing shows pale grey shoulders and a green speculum bordered by white.

Megapodes

THE megapodes are renowned as the 'mound-builders', that incubate their eggs by utilizing the heat of rotting vegetation, the chicks fending for themselves after hatching. The eggs are notably large, with big yolks, and the birds have large feet ('mega-podes') for digging and scraping. Sulawesi's megapodes, of which the endemic

Maleo is unique, in fact do not build mounds, but burrow into the sand or soil and allow solar or geothermal heating to do the job of incubation. They live on the ground, scratching their food from the forest litter, and are rather clumsy fliers. However, one or two megapodes are notable 'island tramp' species, capable of colonizing small islands across the sea.

MALEO
Macrocephalon maleo (50 cm) Frontispiece

Sulawesi's most famous bird is handsome and striking. However, it is shy and can be seen readily only at its communal breeding grounds, which are located on the sandy beaches of coasts or inland rivers, or at sites where there is a thermal heating source. Black volcanic sand is especially favoured. The massive eggs, which are at least four times as heavy as chicken eggs, are in great demand, although their collection is actually illegal. Some fifty breeding sites are known, nearly all in North and Central Sulawesi, but many of these sites are severely threatened. Some additional sites have been discovered recently, but the future of this bird is far from assured. Formerly, the harvesting was controlled by the local communities, but social change arising from population growth and immigration has caused a breakdown of traditional values, resulting in over-exploitation of the eggs. Forest clearance is a major threat, especially where coastal or riverine breeding sites become isolated from the forest hinterland where the birds normally live. Major conservation efforts are being taken, mainly of site management, especially at the Tambun and Tumokang sites in Dumoga-Bone National Park. These are the best locations for the visitor to see Maleo. Although normally shy and silent, the author's first encounter with this fascinating bird was the sound of its extraordinary loud braying voice, coming from three birds perched on middle-storey branches in the alluvial forests of the Lariang River on the west coast.

PHILIPPINE SCRUBFOWL
Megapodius cumingii (30 cm)   Plate 4a

The second megapode on Sulawesi is a much smaller bird, olive-brown and grey, that ranges through the Philippines and Sulawesi, to islands off the coast of North Borneo. Some populations build huge mounds, but in Sulawesi they mostly dig burrows among the roots of forest trees, especially those that are dead and rotting, and their eggs are extremely difficult to find. The call, often heard at night, is a prolonged, quavering whistle.

The Banggai and Sula Islands have their own endemic species, the SULA SCRUBFOWL *M. bernsteinii*; although habitat damage is endangering the future of this bird, recent surveys have shown that it may be still quite common.

The only other 'game birds' (galliformes) in Sulawesi are the RED JUNGLEFOWL *Gallus gallus* (45–60 cm) in the forest, possibly an introduced bird, and the little BLUE-BREASTED QUAIL *Coturnix chinensis* (15 cm) in the grasslands, where also can be found both the RED-BACKED and BARRED BUTTON-QUAILS *Turnix maculosa* and *T. suscitator*. The Red Junglefowl is the ancestor of the domestic chicken, and its crowing call is very similar to the farm bird, but more abbreviated.

Crakes and Rails

THE crakes and rails are secretive wetland birds that creep or swim stealthily about the reed-beds, rice-fields, and swampy pools. They have short, rounded wings and short tails, and tend to fly weakly with dangling legs. Many have a coloured or barred patch beneath the tail, revealed when the tail is cocked. Sulawesi has some distinctive species, in addition to those which occur more widely in Indonesia.

BUFF-BANDED RAIL
Gallirallus philippensis (28 cm)   Plate 4b

This is a common rail of both wet places and dry rice-fields and grassland. It has the typical barred and streaked plumage of many rails, but the distinctive head plumage, with long greyish-white supercilium above the eye, readily serves to identify it. Its range extends from the Philippines to the Pacific and Australasia.

BARRED RAIL
Gallirallus torquatus (30 cm)   Plate 4d

Another common rail in a variety of wet habitats, including forest edge, the Barred Rail is distinguished by having blackish fore-parts, with a pronounced long, white moustachial stripe below the eye, extending to the neck. This species is found also in the Philippines and Irian Jaya.

WHITE-BROWED CRAKE
Poliolimnas cinerea (20 cm)   Plate 4c

This small crake is common and widespread, and it is perhaps more ready than other species to come out into the open. It is quite vocal, being the author of shrill piping sounds, 'weeu, weeu, weeu', often heard in the swamps. A rather plainly coloured bird, there are two prominent white streaks on the dark face.

SNORING RAIL
Aramidopsis plateni (30 cm)   Plate 4e

Sulawesi has two unique, endemic forest rails, but in common with other ground-dwelling birds of the forest, they are extremely elusive. The Snoring Rail is so named from its muted, rather pig-like snoring call, described as 'ee-orrrr'; a second call has been described as a very soft, deep sigh. It has been observed on the ground in dense secondary growth on the forest edge. As Plate 4e shows, the bird appears to be almost tailless, but the bill is rather large. It is somewhat olive above, greyer on the mantle, with a barred abdomen.

BLUE-FACED RAIL
Gymnocrex rosenbergii (30 cm)   Plate 4f

This rail of primary forest is known from both mainland Sulawesi and, at least formerly, Peleng Island. Most of the body is dark grey to black, while the wings and mantle are chestnut. The name is derived from a bluish-white eye patch; the bill is short.

WHITE-BREASTED WATERHEN
Amaurornis phoenicurus (33 cm)   Plate 5b

This is a conspicuous bird with its dark greyish-brown upper-parts, white face, throat, and breast, and chestnut under-tail coverts, and especially because of its curious grunts, chuckles, and repeated 'kru-wak kru-wak' notes. It is common in swampy thickets throughout Sulawesi, and is widespread in South-East Asia and western Indonesia.

ISABELLINE BUSH-HEN
Amaurornis isabellina (26 cm)   Plate 5a

Although one of the waterhen group, this endemic bird is more at home in dry scrub and grassland, though it is not averse to wet places. Another secretive bird, without the shining white fore-parts of the previous species, it is readily identified on the few occasions when good views can be obtained. It has harsh, nasal, and repetitive call-notes, but without the extraordinary cacophony produced by its white-breasted cousin.

COMMON MOORHEN
Gallinula chloropus (33 cm)   Plate 5c

The moorhen is commonly seen close to reeds or rushes on swampy pools, swimming on open water but quickly scurrying to cover when disturbed. Its characteristic sharp, liquid 'prrrt' call is often heard from the depths of the reeds. The yellow-tipped red bill with red frontal shield on the forehead should be noted, and also the oval white patches under the tail.

The Common Moorhen has a nearly world-wide distribution, but it is replaced in Australia by the closely related DUSKY MOORHEN *G. tenebrosa*, which lacks the white line along the flanks. Both species occur together on Sulawesi, although the Dusky Moorhen out-numbers the Common Moorhen on the Tempe lakes.

PURPLE SWAMPHEN
Porphyrio porphyrio (43 cm)   Plate 5d

This much larger rail is common in the bigger reed-beds, in which it hides for much of the time. Occasional short flights above the reeds reveal a heavy, lumpy, bluish-purple bird, with red legs, thick red bill and frontal shield, and white under the tail.

Jacanas

COMB-CRESTED JACANA
Irediparra gallinacea (23 cm)   Plate 5e

Jacanas are very striking, with their long legs and toes, specialized for walking on the floating vegetation of the deeper swamps, earning them the name of 'lily-trotters' or 'lotus birds'. This Australian species ranges west to South-east Kalimantan, but it is rather local in Sulawesi, most often encountered in the south. Its status as a resident has not yet been confirmed. It walks with exaggerated movements, and the plumage, with a large pinkish-red 'comb' in adults, is unmistakable.

Shorebirds

THE many species of brown plovers and waders of coastal and lakeside mud-flats are beyond the scope of a popular book. With just two exceptions, all are migrants from the northern winter. There are thirty-five species on the Sulawesi list, ranging from the tiny stints

(15 cm) to the tall redshanks (28 cm) and godwits (40 cm), and the Whimbrel (43 cm) and curlews (58 cm) with their very long, down-curved bills. The common species inland are the PACIFIC GOLDEN PLOVER *Pluvialis fulva* (25 cm) on short grassland, the COMMON SANDPIPER *Actitis hypoleucos* (20 cm) along rivers and coasts, and the WOOD SANDPIPER *Tringa glareola* (23 cm) in rice-fields.

SULAWESI WOODCOCK
Scolopax celebensis (32 cm)   Plate 5f

One of the resident waders is the endemic Sulawesi Woodcock, which lives in the mountain forests but is very rarely seen. This secretive bird lives on the ground, but makes short flights through or over the canopy at dusk. It has a rather plump body and a very long, straight bill.

The second resident bird is the MALAYSIAN PLOVER *Charadrius peronii* (15 cm), which lives in pairs along sandy beaches. It is a rather sandy-coloured little bird, with white under-parts; in breeding plumage, there is a black bar across the breast.

RED-NECKED PHALAROPE
Phalaropus lobatus (19 cm)

The phalarope is a rather unusual wader that mostly spends the winter swimming offshore, often out on the open sea. It is common in Sulawesi waters between September and May, often seen, for example, by visitors to Bunaken Island in Menado Bay. The Wallacean seas are now known to be a major wintering area for the

Red-necked Phalarope

species. A few birds may also be seen on coastal pools and salt pans. It is a slender, greyish little bird, with a black, needle-like bill; there is a broad white wing bar, and a black smudge through the eye. The name derives from the rufous neck, but this is acquired only in breeding plumage. It has most curious spinning and bobbing movements when feeding, with frequent short hopping flights.

Stilts

WHITE-HEADED STILT
Himantopus leucocephalus (38 cm)

The White-headed Stilt is unmistakable with its very long, pink legs and long, needle-like, black bill, white body, and black wings and hind neck. Its status on Sulawesi is not known, but parties can be seen in most seasons feeding in shallow lakes and coastal pools. This is the Australian cousin of the closely related Black-winged Stilt *H. himantopus* of Europe and Asia.

White-headed Stilt

Thick-knees

BEACH THICK-KNEE
Esacus magnirostris (51 cm)

This widely but very thinly distributed bird occurs throughout Indonesia on remote, sandy beaches, especially on small islands. It is a large, entertaining, but seemingly rather stupid bird, that will trot

Beach Thick-knee

away a short distance along the beach as an observer approaches. Only when pressed will it make a short, looping flight over the sea to a position behind, or further from, the intruder. Greyish-brown above, paler below, it is unmistakable with its massive yellowish bill, large yellow eye, and black-and-white facial markings. In flight, the wing is black-and-white on the upper surface and mainly white beneath. It gives quiet 'wilp' alarm calls, but also piping calls at night, when it is more active.

Terns

TERNS are slender, narrow-winged seabirds with long, pointed bills and long, slightly forked tails. The sea terns that live offshore, diving into the sea or resting on fishing stakes or floating logs, are mostly white, with varying amounts of black on the crown. The BRIDLED TERN *Sterna anaethetus* (42 cm), however, has dark grey-brown upper-parts. The sea terns are somewhat ephemeral visitors to Sulawesi, although some can be seen throughout the year, especially the large GREAT CRESTED *S. bergii* (45 cm) with its greenish-yellow bill, or LESSER CRESTED *S. bengalensis* (38 cm) with orange-yellow bill. The BLACK-NAPED TERN *S. sumatrana* (30 cm) breeds on offshore islets, and recently the LITTLE TERN *S. albifrons* (25 cm) has been discovered breeding on the beach at Morowali in Central Sulawesi.

The terns that are seen on migration over muddy coasts and the adjacent swamps and rice-fields, with rather fluttery flight, swooping to the surface but not diving, are the marsh terns. These are greyish-white in non-breeding plumage, but develop substantial areas of black before departure to their breeding grounds in Australia or Asia; these are the WHISKERED TERN *Chlidonias hybridus* (28 cm) and WHITE-WINGED TERN *C. leucopterus* (25 cm).

Pigeons and Doves

THE pigeons and doves are a major component of the Sulawesi forests. There are no less than twenty-two species of pigeon on the mainland, eight of which are endemic to the Sulawesi and Sula region, with others found on offshore islands. Pigeons and doves may be familiar to most people, but they occur in a wide range of colours—especially the fruit-doves—and in varied habitats.

GREY-CHEEKED GREEN PIGEON
Treron griseicauda (27 cm)   Plate 6a

The green pigeons are here near the eastern limit of their range, and Sulawesi has just two species, this and the PINK-NECKED GREEN PIGEON *T. vernans*. (Recently, the latter has been recorded further east on Halmahera.) Both are common and quite conspicuous, sometimes in flocks, in open, wooded country. Their calls consist of soft chuckling and gurgling notes. Predominantly green, the male Grey-cheeked Green Pigeon has a maroon mantle. The male Pink-necked Green Pigeon has a green mantle, and is distinguished by a broad, pale pink collar and breast, above an orange lower breast band.

SUPERB FRUIT-DOVE
Ptilinopus superbus (22 cm)   Plate 6b

The male fruit-doves, of which Sulawesi has four species, have colourful plumage, but being rather quiet denizens of the middle storey, they are difficult to observe. This beautiful bird has a variety of colours. The male has a purplish crown and orange collar, and a dark breast band separates the light mauve breast from the yellowish-white, green-flanked abdomen. The female is more subdued, but has a dark blue patch on the hind crown. Both sexes have a pale tip to the tail. This species, which extends from the southernmost Philippines to Australia, occurs mainly in the hills, in the forest, and wooded habitats. It has the typical soft 'hoo' notes of the genus, usually a series of such notes that accelerate gently in speed and volume before dying away.

BLACK-NAPED FRUIT-DOVE
Ptilinopus melanospila (23 cm)   Plate 6c

Also common in the same habitat, but more especially in the lowlands, the Black-naped Fruit-dove is distinguished in the male by the grey head and neck with black hind crown and yellow throat. The body is green, with golden-yellow under the tail. The female is mainly green, but has the yellow beneath the tail. The call is a melancholy, repeated 'lu-wu, lu-wu'. Outside Sulawesi, this species occurs on small islands east to Seram and Obi, and west to Java and Lampung Bay.

MAROON-CHINNED (Buff-bellied) FRUIT-DOVE
Ptilinopus subgularis (30 cm)   Plate 6d

This fruit-dove is endemic to lowland forests in the central regions and the northern and eastern arms of Sulawesi, and the Peleng, Banggai, and Sula Islands. The maroon chin on the generally pale grey fore-parts is not very obvious, and the buff spot on the breast is a more useful character. The under-tail coverts are chestnut. It is said to be not very vocal, with a barking 'ko-ko-ko-ko' call.

RED-EARED FRUIT-DOVE
Ptilinopus fischeri (28 cm)   Plate 6e

Another endemic, the Red-eared Fruit-dove is confined to the montane forests of mainland Sulawesi. Also having grey fore-parts, the main features are the red band from the eye to the rear of the head, and dull yellow under-parts. There is some variation in plumage, particularly in the extent of grey and green on the upper-parts. Birds from Gunung Lompobattang in the south have greyish-black upper-parts, only the rump being green, while the under-parts lack colour (*P. f. meridionalis*). It has a distinctive call, 'ooo-wup', rising on the second syllable.

GREEN IMPERIAL PIGEON
Ducula aenea (45 cm)   Plate 6f

The larger, rather plump imperial pigeons are prominent members of the forest and sometimes more open country. The Green Imperial Pigeon is the common form of the lowlands and hills from continental South-East Asia through to Sulawesi and the Sula Islands. It has a pinkish-grey head and under-parts, glossy green upper-parts, a rufous patch on the nape, and red legs. The lack of a pale bar in the tail is an important character. It has a variety of mournful calls, but most typical in Sulawesi is a rolling 'brrrp, brrrp, brrrp' in a series that drops in pitch.

WHITE-BELLIED IMPERIAL PIGEON
Ducula forsteni (50 cm)   Plate 6h

Usually at middle elevations, this is an impressive bird with a green breast sharply demarcated from the white belly, a white head, and white band in the tail. The yellow eye is ringed red. It has a loud, disyllabic booming call, one of the characteristic sounds of the hills of Sulawesi. It is endemic to Sulawesi and the Sula Islands.

GREY-HEADED IMPERIAL PIGEON
Ducula radiata (40 cm)   Plate 6g

In the mountain forests, occasionally lower, the Grey-headed Imperial Pigeon has a greyish head and under-parts and a bronzy mantle. There is a pale bar in the tail. This species is endemic to mainland Sulawesi.

WHITE IMPERIAL PIGEON
Ducula luctuosa (45 cm)

This white pigeon with its black wing and tail feathers should be unmistakable. It occurs widely on forest edges and in the wooded lowlands, and is endemic to Sulawesi and some surrounding islands. It has a typical *Ducula* 'luk-wrroom' call. However, there is a specialist of coasts and small islands that is almost identical, the PIED IMPERIAL PIGEON *D. bicolor* (40 cm). This is smaller, and its creamy under-tail is not mottled with black. The latter bird ranges widely through the oriental archipelago as far as Australia, and will often visit the Sulawesi coast from the offshore islands.

White Imperial Pigeon

SOMBRE PIGEON
Cryptophaps poecilorrhoa (36 cm)   Plate 7a

This large and solitary endemic pigeon is a shy and retiring bird of montane forest, and is either rare, or rarely seen. It has a greyish head and breast, and dark upper-parts and the blackish tail has a pale tip. The eyes and feet are red and the bill is mostly blue-grey. It has a rather noisy, clumsy flight.

SULAWESI BLACK PIGEON (White-faced Cuckoo-dove)
Turacoena manadensis (36 cm)   Plate 7b

This is a beautiful pigeon of the lowland and hill forests, endemic to Sulawesi and the Sula Islands. Plate 7b shows the white head sharply demarcated from the green back of the head and neck, and the darker green body. It has a 3-note call: 'pu, ku-ku', the second note accentuated.

SLENDER-BILLED CUCKOO-DOVE
Macropygia amboinensis (36 cm)   Plate 7c

The reddish-brown cuckoo-doves are represented in Sulawesi by this form, whose range extends east to New Guinea and Australia. This is a very long-tailed brown pigeon often seen in the middle storey of the forest edge, or in swift flight over the canopy, in both the lowlands and hills. The forehead, throat, and under-parts are creamy rufous, the tail is unbarred, and the legs are red. The call is a repeated, up-slurred 'wu-up' at the rate of about 5 in 10 seconds.

RED COLLARED DOVE (Red Turtle-dove)
Streptopelia tranquebarica (23 cm)   Plate 7d

This bird of mainland Asia was discovered in the Palu and Parigi areas of Central Sulawesi only in 1978. It has not been found anywhere else in Sulawesi, or indeed in the Indonesian region, and it has presumably been introduced. It lives in the coconut groves where its rather quiet and rapid 'crr-u-u-u' calls can be heard. It has beautiful soft red and blue-grey plumage.

The SPOTTED DOVE *S. chinensis* (30 cm) is common and widely distributed in open country, and is most often seen as it flies up from roadsides and patches of bare ground, when the black-and-white spotted half-collar and broad white tips to the outer tail feathers can be seen. The call consists of three or four throaty cooing notes.

STEPHAN'S DOVE
Chalcophaps stephani (25 cm)   Plate 7e

It seems strange that two species which are so alike, this and the EMERALD DOVE *C. indica*, should occur in similar habitats on Sulawesi. Probably the former is confined to lowland forest, while the latter occurs more widely in both primary and secondary growth, at all elevations. They are both ground feeders, and are mostly seen as they take off in rapid flight away from the observer, when the green wings and two pale bars on the lower back are very striking. These bars are buff in Stephan's Dove and grey in the Emerald Dove. A pale shoulder patch in the latter (white in the male) is lacking in Stephan's, so is useful in identification when it is seen clearly. Both utter soft 'c-wup' notes, which form a rapid series in Stephan's Dove.

SULAWESI GROUND-DOVE
Gallicolumba tristigmata (32 cm)   Plate 7f

Although quite colourful, the Sulawesi Ground-dove is a very secretive and poorly known bird of the floor of deep forest, mostly in the hills, that is almost never seen. Its call is said to be a quickly repeated, rather smooth and mellow 'ukukukuku'.

NICOBAR PIGEON
Caloenas nicobarica (40 cm)   Plate 7g

Visitors to small, forested islands off the coast of Sulawesi, or any other part of Indonesia, should watch for this island specialist. The slaty-black plumage with a purplish and coppery gloss and, in the adult, the contrasting white tail, are diagnostic. When seen at close range, which is unusual, long 'hackles' are seen arising from the neck. It is a plump, short-tailed pigeon, that scratches around on the floor of the forest, singly or in small groups, but it is extremely wary, every ready to fly up into the trees with a loud clattering of wings. Presumably it is the lack of predators that attracts this pigeon to small islands, and only very rarely are they encountered on the larger islands.

Parrots

THE great family of parrots, familiar to most people, has its centres of distribution in South America (some 120 species), Africa, India, Australia (55 species), and New Guinea (46 species), a fact of interest to students of continental drift. Mainland Sulawesi has nine species, including several endemics, together with others found on its offshore islands. Many are popular as pets, and the trade is increasingly a cause for international concern. They are not everyone's favourite birds, however, and despite gaudy plumage and raucous voice, some are quite difficult to identify. The bright colours can be variable in distribution and difficult to detect in a flock shrieking past at canopy level. They have large heads, and powerful hooked bills for breaking hard fruits and digging out nesting cavities.

YELLOW-CRESTED COCKATOO
Cacatua sulphurea (33 cm)   Plate 8a

Wallacea has four endemic white cockatoos, distinguished mainly by the colour of their crests. As its name implies, the Yellow-crested Cockatoo, which occurs in Sulawesi and Nusa Tenggara, has a yellow crest. The crest is white in Halmahera (*C. alba*), salmon-coloured in Seram (*C. moluccensis*), and pinkish-white in the small Tanimbar Corella (*C. goffini*). They are closely related to the Sulphur-crested Cockatoo *C. galerita* of New Guinea and Australia.

The bird needs no description. It lives in the forest, singly or in small parties, and often descends to farmer's fields early and late in the day, on broad, rounded wings, with loud, raucous calls. It is a sad reflection on the bird trade that it has exterminated this bird from most parts of Sulawesi to the extent that few ornithologists have seen it in recent years, even in the national parks. The Yellow-crested Cockatoo is critically endangered in the wild.

ORNATE LORIKEET
Trichoglossus ornatus (25 cm)   Plate 8b

The Ornate Lorikeet is the common parrot of the lowlands, a very gaudy bird, compact but with rather pointed wings and tail. It occurs in shrill, swift flocks in the forest and wooded country. In flight, the under surface of the wings is red and yellow. The lorikeets are nectar and pollen feeders, but are not averse to taking fruits. Endemic to Sulawesi and its islands, it is closely related to the Rainbow Lorikeet *T. haematodus* of the Lesser Sundas, Seram, New Guinea, and Australia.

YELLOW-AND-GREEN LORIKEET
Trichoglossus flavoviridis (21 cm) Plate 8c

This lorikeet replaces the Ornate in the hills, although there is a considerable zone of overlap. It is another endemic, confined to Sulawesi and the Sula Islands, and lacks the Ornate's gaudy colours. Note that both species have reddish bills. The call is less harsh and grating than the previous species.

Mention must also be made of the RED-AND-BLUE LORY *Eos histrio* (31 cm), a charming and beautifully coloured parrot which is confined to the Sangihe and Talaud Islands, although it is now very rare except on Talaud. Lories differ in having more rounded tails.

SULAWESI HANGING-PARROT
Loriculus stigmatus (15 cm) Plate 8e

The hanging-parrots are so named from their propensity for clambering about with great dexterity, and resting upside-down. They are dumpy little birds, with an almost neckless appearance, generally seen as a flash of deep green in fast, whirring flight, with thin, high calls. The bill of this species is black, but the red forehead of the male can cause an illusion of the bill being red. This is the common lorikeet of flowering trees, endemic to Sulawesi.

RED-BILLED HANGING-PARROT (Tiny Hanging-parrot)
Loriculus exilis (11 cm) Plate 8d

Another endemic, this is much smaller, and differs in having a reddish bill but no red forehead, and no yellowish-orange patch on the upper-back. The call is a series of high-pitched, short, sharp notes. It appears to be nomadic and locally common, but it may be overlooked as the two species are not easy to distinguish.

Sangihe Island has its own endangered endemic, the SANGIHE

HANGING-PARROT *L. catamene*, while the MOLUCCAN HANGING-PARROT *L. amabilis* of Halmahera is found on the Sula and Banggai Islands.

GOLDEN-MANTLED RACQUET-TAIL (Chiming Racquet-tail)
Prioniturus platurus (25 cm)　　　　　　　　　　　　Plate 8f

These parrots are so named from the two long, bare tail shafts, ending in spatulate tips, which extend beyond their quite short tails. They form a small family, confined to the Philippines, Sulawesi, and Buru. This species is common on Sulawesi and the Sula Islands, on woodland edges at all altitudes. Predominantly green, especially the females, it is difficult to view the colours of the male, as mostly these birds are seen in high, swift-flying flocks. However, the grey shoulders are sometimes visible. The rackets vary in length according to age, sex, and the state of moult. It is a noisy bird, constantly calling a high 'keli, keli' in flight.

RED-SPOTTED RACQUET-TAIL (Yellow-breasted Racquet-tail)
Prioniturus flavicans (37 cm)　　　　　　　　　　　　Plate 8g

Perhaps more confined to forest, this endemic parrot has a very restricted distribution. Locally it is common, such as in Dumoga-Bone National Park. The red spot on the crown of the male is not diagnostic, and the alternative name may therefore be more appropriate. The calls are distinctive, however, and the more serious birdwatcher should learn these.

AZURE-RUMPED PARROT (Blue-backed Parrot)
Tanygnathus sumatranus (31 cm)　　　　　　　　　　　Plate 8h

This is a rather solid green parrot, in which the male has a massive red bill, and a blue rump. The female is all green with a white bill. Its scientific name is misleading as the bird is actually endemic to Sulawesi, the Sula Islands, and the Philippines. It is common in all kinds of country in the lowlands, where it causes substantial damage to crops, especially as it commonly feeds at night, when its short, harsh calls are often heard in overhead flight.

Cuckoos

THE true cuckoos are well known as brood parasites, that lay their eggs in the nests of other birds, leaving the host parents to raise the young. They form quite a varied family, and being largely arboreal, are more readily identified by their calls, which are often heard at night as well as through the day. They have rather long, generally barred tails, and a rather hawk-like appearance except for their fine bills. The malkohas and coucals are not parasitic.

RUSTY-BREASTED CUCKOO
Cacomantis sepulcralis (24 cm)   Plate 9a

Sometimes also known as the Indonesian or Oriental Brush Cuckoo (and also previously as *Cuculus variolosus*), this is the most characteristic cuckoo of Sulawesi, being common in both lightly wooded country and the forest margins. It occurs from the Malay Peninsula east to Maluku and the Philippines, with a related species in New Guinea and Australia. The rather long, graduated tail and slightly pointed wings are typical of the cuckoos. Its diagnostic call is a long cadence, a series of plaintive high whistles, 'heet, heet, heet', at a constant speed, falling very slightly in pitch. Sometimes this call can be better rendered as 'psiu, psiu, psiu'. However, both this and the next species also have a rising song that consists of repetitions of 'tay-ta-wi' up the scale, and this alternative call cannot be used for identification.

PLAINTIVE CUCKOO
Cacomantis merulinus (22 cm)   Plate 9b

Sulawesi is the eastern limit of this widespread oriental species, a cuckoo that prefers more open country. Very similar in plumage to the previous species, the two cuckoos can best be identified by their cadence songs. In the Plaintive Cuckoo, this begins with three or four slow 'heet' notes, but these then break into a rapid cadence of about eight notes down the scale.

In the hill forests, an observer should listen for the endemic SULAWESI HAWK-CUCKOO *Cuculus crassirostris*. A canopy bird that is rarely seen, its call is a beautiful series of two or three notes,

DRONGO CUCKOO
Surniculus lugubris (23 cm)   Plate 9d

'ka-ku', or 'ka-ka-ku' (the second or third note is lower), with other variants such as a musical 'dong-dong'.

This is a black drongo-like cuckoo of wooded terrain. It is readily recognized by its rather human whistle of seven to nine notes rising the scale (generally two notes more than the call from this species in the western regions of Indonesia). If the bird is not calling, the bars in the tail, and the fine bill, should serve to distinguish this from a drongo (see p. 71).

GOULD'S BRONZE CUCKOO
Chrysococcyx russatus (15 cm)   Plate 9c

Bronze Cuckoos are small and inconspicuous arboreal birds that are often heard but rarely seen. The calls are distinctive, but soft and easily overlooked until they are known. One call is a tinkling cadence of some five to eight rapid, high, clear 'tee-tee-tee' notes, lasting about one second, and the second call is a slightly longer, tinkling trill on one pitch. It is a colourful bird, with its bronze-green upper-parts, with a rufous wash in places, barred under-parts, and red eye. The host species probably includes the Flyeater (see p. 56). It is found commonly from wooded areas in the hills to gardens in towns such as Palu, but it has yet to be recorded from the Minahassa region. It is one of a complex group of species that extends from Australia to Thailand.

BLACK-BILLED KOEL
Eudynamys melanorhyncha (43 cm)   Plate 9e

This koel is a common bird of forests and wooded areas, especially in the lowlands. The male is all black, slightly glossed, while the female is generally more rusty and barred. Both sexes have a black bill. This bird has the same calls as the Asian Koel *E. scolopacea* of Java and the oriental region, a rising series of 'ko-el' calls in the male, and a crescendo of excited bubbling calls from both sexes. However, the most characteristic call in Sulawesi, probably from male birds in the forest canopy and often heard at night, is a fast, rising and falling 'wrr-wrr-wrr-wrr-wrr-wrr'. This is one of the mournful calls that give the Sulawesi forests their melancholy quality. The Short-crested Myna (see p. 70) has been recorded as a host species.

CHANNEL-BILLED CUCKOO
Scythrops novaehollandiae (60 cm)   Plate 9i

This huge cuckoo is a remarkable bird. Usually seen in twos or threes, occasionally small parties, it is unmistakable with its grey plumage, huge, curved, straw-coloured bill, bare red skin round the eye, and 'maltese cross' silhouette in flight (elongated fore-parts and tail, and long wings). It flies with slow, shallow wing-beats and glides. The most notable feature is the extraordinary loud, raucous, and even ugly call, that drowns out all other sounds in the forest. Although resident, numbers may be augmented from Australia during the southern winter. It lays its eggs in the nests of crows.

YELLOW-BILLED MALKOHA
Rhamphococcyx calyorhynchus (45 cm)   Plate 9f

This endemic species of Sulawesi, and another in the Philippines, are the only malkohas found east of Wallace's Line. The Yellow-billed Malkoha is a common bird, that flops rather heavily about in the middle storey of forest and scrubby woodland. The plumage shown in Plate 9f is unmistakable. The calls consist of mewing notes and a characteristic rattling call that rises and falls in pitch.

BAY COUCAL
Centropus celebensis (45 cm)   Plate 9h

Coucals are large, heavy birds of the undergrowth. The endemic Bay Coucal is a brown bird of the understorey in all kinds of forest, in twos or threes, but far more often heard than seen. The call consists of a series of mournful 'boob-boob-boob' notes, very similar to those of the Greater Coucal *C. sinensis* of the Sundaic region, but variable in speed and pitch; the full series rises slightly before falling and then rising again to a tapering finish. More usually, however, the series is truncated abruptly after a few notes. One bird calling will almost invariably set off other birds, both those in the same group and others near by, creating a most melancholy effect.

LESSER COUCAL
Centropus bengalensis (40 cm) Plate 9g

This coucal is common in scrubby growth and grassland in the lowlands, in Sulawesi and throughout the Sundaic and Wallacean region. It is a streaky, dark brown to black bird with chestnut wings. It has both hollow 'but-but' notes and a more staccato, unmusical call.

Owls

MANY people have a superstitious fear of these predators of the night, but they are fascinating birds, eagerly sought by ornithologists. Mainland Sulawesi has six resident species, five of which are endemic, and little is known about most of them.

SULAWESI OWL (Rosenberg's Owl)
Tyto rosenbergii (45 cm) Plate 10a

The world-wide Barn Owl *T. alba* (which occurs on several Indonesian islands) is replaced on Sulawesi by this endemic species, but it occupies the same open country habitat, and it is probably common. It also occurs on Sangihe and the Peleng Islands. Plate 10a shows a much darker bird than the Barn Owl, much spotted, and with five bold black bars on the tail. It probably roosts by day in large trees, emerging at dusk to hunt on round, silent wings, occasionally uttering its eerie, long shrilling call, 'skkeear'. This call is almost identical to that of the Barn Owl.

Sulawesi has another endemic 'barn owl', the MINAHASSA (or Sulawesi) MASKED OWL *T. inexspectata*, a forest owl known at time of writing from just eleven specimens and two sightings. All but one of these were in North Sulawesi, but the discovery of a dead bird in Lore Lindu National Park in 1980 indicates a wider range. The Sula Islands have their own endemic, the TALIABU MASKED OWL *T. nigrobrunnea*, which is known only from a single female specimen, although there has been a recent sighting.

Finally in this group, the EASTERN GRASS-OWL *T. longimembris* (35 cm) has also been recorded in Sulawesi, but little is known about its distribution on the island.

SULAWESI SCOPS OWL
Otus manadensis (20 cm)   Plate 10b

Scops Owls are small owls with pronounced ears tufts (upright feathers on the sides of the head). This is a common endemic of Sulawesi. Its rather human whistled rising 'pewit' call, a clear note uttered at intervals, can often be heard in both forest and more open woodland, in the lowlands and hills. This bird is confined to mainland Sulawesi, and is replaced on the offshore islands by the more widespread MOLUCCAN SCOPS OWL *O. magicus* (25 cm). This has a call consisting of a gruff single note uttered at short intervals.

SPECKLED BOOBOOK
Ninox punctulata (24 cm)   Plate 10c

The boobooks, otherwise known as hawk-owls, lack the flat facial disc of most owls, giving them a rather dumpy, hawk-like appearance. There are two endemic species on Sulawesi. The Speckled Boobook is common in a variety of forest and more open, wooded habitats in the lowlands and low hills. The call has been described as a series of 'toy' sounds, with a rather sibilant quality, slightly rising in pitch and accelerating. However, the author has heard a variety of owl calls in the Sulawesi forests, and much more work is needed to clarify the calls of the various endemic species.

OCHRE-BELLIED BOOBOOK
Ninox ochracea (24 cm)   Plate 10d

The Ochre-bellied Boobook is more confined to forests, in the lowlands and up to 1780 metres, recorded from North, Central and South-east Sulawesi, and Buton Island, but little is known about it. The call is described as a series of rather guttural notes on one pitch, at a rate slightly faster than one per second, but tending to reduce speed and evolve into a slower series of double notes.

1. (a) Pied Heron, p. 5. (b) Striated Heron, p. 5. (c) Black-crowned Night-heron, p. 6. (d) Rufous Night-heron, p. 5. (e) Cinnamon Bittern, p. 6. (f) Yellow Bittern, p. 6. (g) Black Bittern, p. 6.

2. (a) Osprey, p. 8. (b) Sulawesi Serpent-eagle, p. 11. (c) Spot-tailed Goshawk, p. 12. (d) Vinous-breasted Sparrow-hawk, p. 12. (e) Spotted Harrier, p. 12. (f) Spotted Kestrel, p. 12. (g) Barred Honey-buzzard, p. 11. (h) Sulawesi Hawk-eagle, p. 11.

3. (a) Red-throated Little Grebe, p. 1. (b) Wandering Whistling-duck,
p. 13. (c) Spotted Whistling-duck, p. 13. (d) Sunda Teal, p. 13.
(e) Pacific Black Duck, p. 14. (f) Garganey, p. 14.

4. (a) Philippine Scrubfowl, p. 16. (b) Buff-banded Rail, p. 16. (c) White-browed Crake, p. 17. (d) Barred Rail, p. 17. (e) Snoring Rail, p. 17. (f) Blue-faced Rail, p. 17.

5. (a) Isabelline Bush-hen, p. 18. (b) White-breasted Waterhen, p. 18.
(c) Common Moorhen, p. 18. (d) Purple Swamphen, p. 19.
(e) Comb-crested Jacana, p. 19. (f) Sulawesi Woodcock, p. 20.

6. (a) Grey-cheeked Green Pigeon, p. 23. (b) Superb Fruit-dove, p. 24.
(c) Black-naped Fruit-dove, p. 24. (d) Maroon-chinned Fruit-dove,
p. 24. (e) Red-eared Fruit-dove, p. 25. (f) Green Imperial Pigeon,
p. 25. (g) Grey-headed Imperial Pigeon, p. 26. (h) White-bellied
Imperial Pigeon, p. 25.

7. (a) Sombre Pigeon, p. 26. (b) Sulawesi Black Pigeon, p. 27.
(c) Slender-billed Cuckoo-dove, p. 27. (d) Red Collared Dove, p. 27.
(e) Stephan's Dove, p. 28. (f) Sulawesi Ground-dove, p. 28.
(g) Nicobar Pigeon, p. 28.

8. (a) Yellow-crested Cockatoo, p. 29. (b) Ornate Lorikeet, p. 29.
(c) Yellow-and-Green Lorikeet, p. 30. (d) Red-billed Hanging-parrot,
p. 30. (e) Sulawesi Hanging-parrot, p. 30. (f) Golden-mantled Racquet-
tail, p. 31. (g) Red-spotted Racquet-tail, p. 31. (h) Azure-rumped
Parrot, p. 31.

9. (a) Rufous-breasted Cuckoo, p. 32. (b) Plaintive Cuckoo, p. 32.
(c) Gould's Bronze Cuckoo, p. 33. (d) Drongo Cuckoo, p. 33.
(e) Black-billed Koel, p. 33. (f) Yellow-billed Malkoha, p. 34.
(g) Lesser Coucal, p. 35. (h) Bay Coucal, p. 34. (i) Channel-billed Cuckoo, p. 34.

10. (a) Sulawesi Owl, p. 35. (b) Sulawesi Scops Owl, p. 36. (c) Speckled Boobook, p. 36. (d) Ochre-bellied Boobook, p. 36. (e) Great-eared Nightjar, p. 37. (f) Sulawesi Nightjar, p. 37. (g) Savanna Nightjar (central region form), p. 37.

11. (a) Collared Kingfisher, p. 40. (b) Sacred Kingfisher, p. 41. (c) Black-billed Kingfisher, p. 42. (d) Sulawesi Dwarf Kingfisher, p. 42. (e) Common Kingfisher, p. 41. (f) Blue-eared Kingfisher, p. 41. (g) Green-backed Kingfisher, p. 43; (gi) form in east and south. (h) Scaly-breasted Kingfisher (Minahassa form), p. 43; (hi) form in central regions. (i) Lilac-cheeked Kingfisher, p. 43.

12. (a) Blue-tailed Bee-eater, p. 44. (b) Rainbow Bee-eater, p. 44.
(c) Purple-bearded Bee-eater, p. 45. (d) Purple-winged Roller, p. 45.
(e) Common Dollarbird, p. 46.

13. (a) Sulawesi Hornbill, p. 46. (b) Knobbed Hornbill, p. 46. (c) Ashy Woodpecker (form in north), p. 47; (ci) form in rest of Sulawesi. (d) Sulawesi Woodpecker, p. 47. (e) Red-bellied Pitta, p. 47. (f) Hooded Pitta, p. 48.

14. (a) White-rumped Cuckoo-shrike, p. 50. (b) Pygmy Cuckoo-shrike, p. 51. (c) Caerulean Cuckoo-shrike, p. 50. (d) Pied Cuckoo-shrike, p. 50. (e) Sulawesi Cicadabird, p. 51. (f) Sulawesi Triller, p. 51.

15. (a) Great Shortwing, p. 52; (ai) form of female in north. (b) Sulawesi Thrush, p. 53. (c) Island Thrush, p. 54. (d) Red-backed Thrush, p. 53.

16. (a) Sulawesi Babbler, p. 54. (b) Clamorous Reed-warbler, p. 56.
(c) Golden-headed Cisticola, p. 57. (d) Malia, p. 55. (e) Sulawesi Leaf-warbler, p. 56. (f) Mountain Tailorbird, p. 57. (g) Chestnut-backed Bush-warbler, p. 56. (h) Geomalia, p. 55.

17. (a) Rufous-throated Flycatcher, p. 58. (b) Mangrove Blue Flycatcher, p. 58. (c) Blue-fronted Blue Flycatcher, p. 58. (d) Rusty-bellied Fantail, p. 60. (e) Citrine Flycatcher, p. 59. (f) Black-naped Monarch, p. 60. (g) Island Flycatcher (form in north, central, and south-eastern regions), p. 59. (h) Sulphur-bellied Whistler, p. 61. (i) Maroon-backed Whistler, p. 61.

18. (a) Grey-sided Flowerpecker, p. 62. (b) Crimson-crowned Flowerpecker, p. 62. (c) Yellow-sided Flowerpecker, p. 62. (d) Black Sunbird, p. 63; (di) male in north region. (e) Crimson Sunbird, p. 63. (f) Crimson Myzomela, p. 65. (g) White-eared Myza, p. 66.

19. (a) Lemon-bellied White-eye, p. 64. (b) Black-fronted White-eye, p. 64. (c) Lemon-throated White-eye, p. 65. (d) Streak-headed Darkeye, p. 65. (e) Pale-headed Munia, p. 66. (f) Black-faced Munia, p. 67. (g) Scaly-breasted Munia, p. 67. (h) Sunda Serin, p. 67.

20. (a) Asian Glossy Starling, p. 68. (b) Finch-billed Myna, p. 69.
(c) White-vented Myna, p. 69. (d) White-necked Myna, p. 69;
(di) form in north and east regions. (e) Short-crested Myna, p. 70.
(f) Fiery-browed Myna, p. 70. (g) Black-naped Oriole, p. 70.

Nightjars

GREAT-EARED NIGHTJAR
Eurostopodus macrotis (37 cm)   Plate 10e

These rather hawk-like birds that hunt particularly as dusk turns to night have beautifully mottled, cryptic plumage that enables them to lie up undetected by day, on branches or on the ground. This common species emerges from the forest and woodland at dusk to hawk for insects over the canopy or clearings. It is recognizable by its rather long wings and tail in silhouette, and its 'taweeo' or 'pit-weeu' calls (softer than the 'tok, tedau' of its Sundaic cousin, the Malaysian Eared Nightjar *E. temminckii*). This species, which ranges from India across mainland Asia to the Philippines, Sulawesi, and the Sula Islands, is known elsewhere in Indonesia only from Simeuleuwe Island off the west coast of northern Sumatra.

The Minahassa peninsula also has the mysterious SATANIC NIGHTJAR *E. diabolicus*, which is known definitely only from a single female specimen. It may possibly be already extinct.

SAVANNA NIGHTJAR
Caprimulgus affinis (23 cm)   Plate 10g

This short-tailed nightjar is a common bird of very open country, the sparse scrub at the back of beaches, and sometimes towns. The white outer tail feathers should be noted as it flits around the roadsides at dusk. The bird illustrated in Plate 10g is the rather pale *C. a. propinquus* of Central Sulawesi; the nominate form in the south has more black in the plumage. Its querying 'schwick' calls at night immediately identify it, and can be heard, for example, at Palu golf course, or in downtown areas of Ujung Pandang.

SULAWESI NIGHTJAR
Caprimulgus celebensis (30 cm)   Plate 10f

The Sulawesi Nightjar was formerly considered to be a subspecies of the Large-tailed Nightjar *C. macrurus*, which is a familiar bird of the oriental region, with its monotonous 'chonk chonk' calls.

Subsequently considered to be a subspecies of the Philippine Nightjar *C. manillensis*, it is now treated as a full endemic species. It is poorly known, and apparently confined to northern Sulawesi and the Sula Islands. A bird of wooded regions, in North Sulawesi it is reported to occur in coastal scrub and mangroves. The call is described as a rapid series of eight or nine 'chuck' notes, trailing off and lasting little more than a second.

Swifts

SWIFTS are aerial insect-feeders identified by their long, slender, curved wings and predominantly dark or black plumage, that spend almost the entire day in the air (some even sleep on the wing). The tails are often forked. There are three or four groups of species, and the small swiftlets and the large needletails are notoriously difficult to identify. The swiftlets include those that build the famous edible nests (composed of saliva) in caves or houses, but scientists still await confirmation that such nests are found on mainland Sulawesi. Some swiftlets have the ability to navigate in caves by echo location.

GLOSSY SWIFTLET
Collocalia esculenta (10 cm)

Swiftlets are seen everywhere in Sulawesi. They do not have such long wings as other swifts, and so have a more stocky silhouette despite their small size. This is a difficult group for the beginner, and the taxonomy is complex, but this species can be readily identified by its glossy bluish-black upper-parts and whitish under-parts.

A very common, uniformly brown species is likely to be the UNIFORM SWIFTLET *Aerodramus vanikorensis*, while the MOLUCCAN SWIFTLET *A. infuscatus* has noticeably pale under-parts and a pale rump.

Glossy Swiftlet

Two apparently new arrivals in South Sulawesi are the LITTLE SWIFT *Apus affinis* (15 cm) and the ASIAN PALM-SWIFT *Cypsiurus balasiensis* (13 cm). The Little Swift has a solid black body, rather long wings, and a very prominent white rump. Noisy twittering flocks are found in some towns, or near road bridges; in the north, there is one report of birds breeding on a church at Kotamobagu. The delicate Palm-swift is identified by its long, slender, curved wings and rather long tail; it is common around palm trees in the open country of the south, and has also been reported from Tinanggea in the south-east.

PURPLE NEEDLETAIL
Hirundapus celebensis (23 cm)

The needletails are so named from their short, rather stubby tails that terminate in short spines. They are large swifts, capable of very fast flight that creates a whooshing sound at close range. This species, which is endemic to Sulawesi and the Philippines, is related to the Brown-backed Needletail *H. giganteus* of the oriental region. It is all black with a bluish gloss, a grey spot on the cheek, and white under-tail coverts. It is often reported in flocks preceding a thunderstorm.

Needletails seen during the northern winter, with a whitish throat, rump and under-tail coverts, are likely to be the migrant WHITE-THROATED NEEDLETAIL *H. caudacutus* (20 cm), which passes only briefly through Sulawesi.

Purple Needletail

GREY-RUMPED TREE-SWIFT
Hemiprocne longipennis (20 cm)

Unlike the true swifts, tree-swifts habitually perch on the open branches of trees. They are found in the more wooded areas of the lowlands, and are distinguished by their very long, needle-like forked tails, and very long, stiff, crescentic wings. This common species ranges from the Malay Peninsula through western Indonesia to Sulawesi. It has a short crest, glossy greenish-grey upper-parts, a paler grey rump and under-parts, and dark wings and tail. The male has a chestnut cheek patch. Large parties sometimes gather to roost in a single tree.

Grey-rumped Tree-swift

Kingfishers

THE kingfishers, with their cheerful colours and large, dagger-like bills, are familiar to most people. The larger species can be quite noisy. They are not all fish-eaters, however, and they occur in a variety of habitats. Those that live in the forest can be extremely difficult to find. Sulawesi is famous for having no less than ten resident species, of which six are endemic. However, four of the endemics are inconspicuous forest dwellers.

COLLARED KINGFISHER
Halcyon chloris (24 cm)   Plate 11a

The Collared Kingfisher is the most conspicuous and noisy member of the family. It occurs throughout Sulawesi in open country, both on the coasts and inland, where it is readily identified by its clean blue and white plumage and loud, almost irritating, 'kek-kek, kek-kek-kek' calls. It also has a softer, querying call, 'pit-pree, pit-pree'. It feeds on large insects, and lives in the outer branches of the trees, but will sometimes descend to the ground, especially along the beach, to look for crustaceans. It is a common bird throughout South-East Asia to Australia.

The Talaud Islands have their own endemic TALAUD KINGFISHER *H. enigma*, which is similar to the Collared Kingfisher but with a smaller bill. Both species occur there, the latter species preferring more secluded and wooded sites.

SACRED KINGFISHER
Halcyon sancta (20 cm)   Plate 11b

This very similar bird is a migrant from Australia during the southern winter, from about April to October. It is found in the same type of terrain as the Collared Kingfisher, where it is nearly as common. It is smaller and duller, appearing slightly dirty in contrast, with a darkish tinge above and buff tinge below. Generally silent in its winter quarters, it is a more familiar and approachable bird than its larger cousin, favouring lower perches, along the coast and in village gardens.

COMMON KINGFISHER
Alcedo atthis (18 cm)   Plate 11e

This kingfisher, the common species of Europe, occurs widely along rivers and coasts in Sulawesi. It is a small kingfisher, blue above and bright rufous below, with a white patch on the side of the upper neck. The resident subspecies on Sulawesi does not have rufous ear coverts. A brilliant blue stripe runs down the back, visible as the bird flies away from the observer in its darting flight, the most common view. The call is a single, shrill note. It is a common resident, but there is also a migrant subspecies from the north, which has the more greenish-blue upper-parts and distinctive rufous ear coverts that characterize this species elsewhere in its range.

BLUE-EARED KINGFISHER
Alcedo meninting (15 cm)   Plate 11f

The Blue-eared Kingfisher is a Sundaic species that extends east to Sulawesi and the Sula Islands. Elsewhere it can be quite readily distinguished from the Common Kingfisher, but the darker resident subspecies of the latter bird in Sulawesi lacks rufous ear coverts. The two resident *Alcedo* kingfishers on Sulawesi are very similar, and even experienced bird-watchers have difficulty in distinguishing them. It is believed that the calls are a useful guide to identity. The Blue-eared Kingfisher favours more enclosed rivers and streams in the forest.

BLACK-BILLED KINGFISHER
Pelargopsis melanorhyncha (37 cm)   Plate 11c

This large kingfisher, which is endemic to Sulawesi and the Sula Islands, is Sulawesi's equivalent of the Stork-billed Kingfisher *P. capensis* of the Sundaic region. It has the same massive bill, and very loud, cackling and wailing but not wholly unmusical calls, and it occupies the same habitats of the more remote coasts, mangroves, estuaries, and occasionally inland swamps or even wooded river valleys. There the likeness ends, however, for it is a pale bird, creamy in colour with brown back, wings, and tail, with a greenish gloss; the bill is dark. It is rather scarce, and appears to be now absent from all of South Sulawesi south of Mamuju.

SULAWESI DWARF KINGFISHER
Ceyx fallax (14 cm)   Plate 11d

The other endemic kingfishers are all inconspicuous forest dwellers. The Dwarf Kingfisher is usually seen as a tiny flash of reddish colour in the understorey of the forest, often far from streams. The mantle and wings are reddish and the rump is blue. The bill is scarlet, and there are black spots on the crown. The under-parts are a rusty yellow.

This species is confined to Sulawesi and Sangihe Island. The Sula Islands have the lovely VARIABLE DWARF KINGFISHER *C. lepidus*, which ranges from the Philippines through Maluku to New Guinea. This bird is a beautiful deep blue above and rufous below.

GREEN-BACKED KINGFISHER (Blue-headed Wood-kingfisher)
Actenoides monachus (31 cm) Plate 11g

A denizen of the dark forest, where it moves around quietly and may be partly crepuscular, this bird is confined to mainland Sulawesi. The whirring of its wings in flight may reveal its presence, but it is difficult to get a view. It is quite common in lowland forest, where its call can be heard in the pre-dawn and dusk, a mournful 'huuuuEEEu', uttered at short intervals, the first part rising and the last syllable higher pitched. It is a rather heavily built, heavy-billed kingfisher. The nominate form of North and Central Sulawesi is olive above with a blue-green rump and tail, dark blue crown, and red bill. The under-parts are rusty, whiter on the throat. *A..m. capucinus* on the eastern, south-eastern, and southern peninsulas has all the blue parts on the head replaced by black (see Plate 11gi).

SCALY-BREASTED KINGFISHER
Actenoides princeps (24 cm) Plate 11h

This smaller kingfisher replaces the previous species in the higher hill forests. The pale feather edges give a scaly appearance to the plumage. The nominate form of Minahassa has a yellowish-brown bill, while elsewhere the bill is red and the collar is more rusty. The call is reported to be softer and more melancholy, rolled on the first syllable and rising on the second.

The RUDDY KINGFISHER *Halcyon coromanda* (25 cm) is largely a coastal bird, although in Sulawesi it extends quite far inland in lowland forest, usually near rivers. It is violet rufous above, paler below, with a shining pale blue rump and red bill. The call consists of two or three very sharp descending notes, sometimes only a single note, or up to six in a run; this song is often given at dusk or before dawn.

LILAC-CHEEKED KINGFISHER
Cittura cyanotis (28 cm) Plate 11i

This is an endemic genus, confined to the lowlands of Sulawesi and the Sangihe Islands, distinguished by its broad, flat bill. It also lives in the interior of the forest and is difficult to see. The call is described as a fast, explosive 'kukukuku', unlike that of any other kingfisher, although the call on a recording by David Gibbs consists of four rather plaintive descending notes, repeated at three-second intervals. In view of the almost total forest clearance on the Sangihe Islands, it seems unlikely that the subspecies *sanghirensis* can still be extant.

Bee-eaters

BEE-EATERS are graceful, slender, and colourful birds that feed by taking insects on the wing, either in continuous flight or by making short sallies from a perch. Some species, notably in Africa, regularly feed on bees. The bill is long, fine, and slightly curved, and the tail is rather long, often with central streamers.

BLUE-TAILED BEE-EATER
Merops philippinus (30 cm)　　　　　　　　　　　　　　　　Plate 12a

In most parts of Indonesia, this widespread species is a winter visitor from the north. However, it is also quite a common resident in the drier regions of Sulawesi, such as the Palu area and the south. A rather large bee-eater, it is greenish-blue above, with a shining blue back and tail, and a yellow throat bordered below by a brown band. The two central tail feathers are elongated. It feeds in continuous flight, when the under-wings reveal a coppery gloss. The call is a short, rolling, trilled note: 'kwilp, kwilp'. It breeds communally in holes excavated in sandy river banks and road cuttings.

RAINBOW BEE-EATER
Merops ornatus (25 cm)　　　　　　　　　　　　　　　　Plate 12b

This smaller bee-eater is a common migrant from Australia from April to September, in parties or sometimes quite large flocks. It is similar to the Blue-tailed Bee-eater, but slightly more colourful, although the bordering breast band is black rather than brown, and the tail is black. The flight feathers have a coppery gloss above and below, and the crown is also bronzy. The call note is higher pitched. In flight, the two species are not always readily distinguished. It feeds in continuous flight, but is more willing than the previous species to feed by making short sallies from exposed perches.

PURPLE-BEARDED BEE-EATER
Meropogon forsteni (26 cm)   Plate 12c

The Purple-bearded Bee-eater is one of the endemics in Sulawesi's unique avifauna that so delight bird-watchers. It is a rather uncommon bird of the forests, and confined to the north, central, and south-eastern arms of the island. It has a purplish head and under-parts, which appear almost black in a poor light, and grass-green upper-parts, with chestnut on the nape, and the central tail feathers are moderately elongated. It is a bird of the forest, but prefers perches at the edge of small clearings, from where it makes short feeding sallies. It has a rather upright, thick-necked stance when perched, wagging its tail up and down almost like a metronome. The call is a short, shrill but quiet double note: 'psit-psit'.

Rollers

ROLLERS are rather stout, thick-headed and short-tailed birds that are typically seen atop a prominent perch, from whence they make short sallies in the air or to the ground in chase of insects. The name is derived from the acrobatics performed in display flight. They have harsh, croaking calls.

PURPLE-WINGED ROLLER
Coracias temminckii (32 cm)   Plate 12d

This common endemic of open country and wooded areas in Sulawesi is an isolated relative of the widespread Indian Roller *C. benghalensis* of mainland Asia. It is readily identified by its stocky build, and purplish-blue colour on both body and wings, with a shining bright turquoise crown and rump.

COMMON DOLLARBIRD
Eurystomus orientalis (30 cm) Plate 12e

At a distance, the Common Dollarbird looks very similar to the Purple-winged Roller, but at close range, it is seen to have a shorter bill, red in colour, and in flight it has conspicuous circular silvery patches in the wings. The Common Dollarbird is a widespread species of well-wooded terrain and forest clearings.

Hornbills

THESE huge forest birds with their massive bills are striking and impressive even to the least interested observer. Sulawesi has two species, both common and both endemic.

KNOBBED HORNBILL
Rhyticeros cassidix (male 100 cm, female 88 cm) Plate 13b

The larger of the two hornbills, this bird has a black body and wings and a white tail. There is a huge casque on the bill, red in the male and yellow in the female. The bill itself is yellow in both sexes, and there is a blue pouch on the throat with a dark bar across it. The male has a chestnut cap, and white neck and upper breast, often stained yellow. The neck is black in the female. They fly around and above the canopy in small, loose parties, but occasionally in gatherings of fifty or more, uttering loud, goose-like, braying notes, while the wings in flight make a noise like a steam engine.

SULAWESI HORNBILL (Sulawesi Dwarf Hornbill)
Penelopides exarhatus (male 60 cm, female 50 cm) Plate 13a

The Sulawesi Hornbill is much less conspicuous, being smaller and tending to keep to the middle storey, inside the forest, often in small garrulous parties. The male has a whitish face and neck, though this is generally stained yellow; there is a black dorsal line on the crown and back of the neck. The casque is small and reddish. It has chattering 'kirra, kirra' calls that sometimes break into falsetto.

Woodpeckers

THE woodpeckers are here at the very eastern limit of their range, with just two species in Sulawesi, both endemic.

ASHY WOODPECKER
Mulleripicus fulvus (38 cm)	Plate 13c

The Ashy Woodpecker is common in forest and wooded terrain. It is a rather elongated woodpecker, greyish above and variably creamy to light rufous below, with a dark bill, a rather prominent white eye, and a red crown in the male. It has a whinnying call, consisting of some eight to twelve rapid notes on one pitch. It also drums on wood as a form of territorial 'vocalization'. A gentle tapping will often first reveal a bird's presence as it digs out grubs from the bark of a tree. In the male of the nominate form in North Sulawesi, the fore crown is red, but the red extends to the entire crown in *M. f. wallacei* elsewhere on the island.

SULAWESI WOODPECKER
Dendrocopos temminckii (15 cm)	Plate 13d

This small and stumpy woodpecker is mainly olive with buff bars above, paler and streaked below, with a striped pattern on the side of the head. The male has a red band around the back of the neck. It is found in any wooded habitat, avoiding closed forest, but it is easily overlooked.

Pittas

RED-BELLIED PITTA (Blue-breasted Pitta)
Pitta erythrogaster (17 cm)	Plate 13e

Pittas are very colourful, stumpy, almost tailless birds that live mainly on the forest floor and in the understorey. Although the colours are striking, they tend to be very elusive birds. The Red-bellied Pitta is more readily seen than many pittas, though only by a patient bird-watcher. Its greyish-blue breast and scarlet belly are diagnostic. The very distinctive call is a mournful, long, double whistle, the second note generally rising, or rising then falling. Its range extends from the Philippines to Australia, but recently the subspecies occurring on the Sula and Banggai Islands has been separated as the SULA PITTA *Pitta dohertyi*.

HOODED PITTA
Pitta sordida (20 cm)　　　　　　　　　　　　　　　　　Plate 13f

The Hooded Pitta is a rather uncommon resident in the hill forests of northern Sulawesi. The author has heard it on Sangihe Island and in a wooded valley near Palu. The entire head region is black, and the body is mostly green, except for blue on the rump and scarlet on the belly. There is a light blue patch in the wing coverts, but Sulawesi birds lack the white patch in the wing that this bird usually displays. The call reveals the bird's presence, a double 'wilp-wilp' uttered at about three- to five-second intervals.

Swallows

PACIFIC SWALLOW
Hirundo tahitica (14 cm)

Most people are familiar with the swallows, aerial insect-eaters that feed on the wing and rest communally on telegraph wires or other prominent perches. They have a swooping flight, with rather curved wings and slightly forked tails, and most species have blue upper-parts. The Pacific Swallow is a common resident in villages throughout Sulawesi, building mud nests on the rafters or eaves of houses. It has blue upper-parts, a chestnut throat, and a dirty white belly.

Pacific Swallow

　The BARN SWALLOW *H. rustica* is a very common migrant during the northern winter, some arriving as early as July. It has cleaner white under-parts, separated from the chestnut throat by a black band, and the adult has greatly elongated outer tail feathers. However, many wintering birds are immature, and the two species are then not so easy to tell apart.

Wagtails and Pipits

YELLOW WAGTAIL
Motacilla flava (18 cm)

The wagtails are unmistakable, hunting for insects by walking or running along the ground, constantly wagging their tails up and down for balance. They visit Sulawesi during the northern winter. The Yellow Wagtail is found in rice-fields and open swamps in flocks that occasionally number thousands. It is olive-green above, but often with a grey crown, and yellow below, with white outer tail feathers, and a wheezy 'tseep' call.

The GREY WAGTAIL *M. cinerea* (19 cm) has a longer tail and the upper-parts are greyer. Its call is a sharper 'chizik'. It is a solitary bird, most often seen along forest roads although migrant birds will occur almost anywhere.

Yellow Wagtail

The pipits are streaky brown birds that lack such long tails. RICHARD'S PIPIT *Anthus novaeseelandiae* (18 cm) is resident in pasture and cultivation in the south.

Cuckoo-shrikes and Trillers

CUCKOO-SHRIKES are medium-sized birds, mainly varying shades of grey but with some black or white in the plumage. They are quite prominent and often noisy birds of the canopy or middle storey, but their taxonomy is complex. The five species on mainland Sulawesi are all more or less endemic. The trillers are smaller and more familiar birds, white-and-black in the male, white-and-grey or brown in the female.

WHITE-RUMPED CUCKOO-SHRIKE
Coracina leucopygia (25 cm)   Plate 14a

The White-rumped Cuckoo-shrike is a common bird in open wooded areas and cultivation in the lowlands. It is medium grey, with a white supercilium, white rump and belly, and black tail and flight feathers. It is seen in twos or threes or in small, noisy parties, with 'shraik, shraik' and other harsh calls.

PIED CUCKOO-SHRIKE
Coracina bicolor (male 32 cm, female 28 cm)   Plate 14d

This is more distinctive, with the males glossy black above, with a white rump, and white below. The female has most of the black replaced by grey, which also extends on to the breast, and only the wings and tail are black. It is generally encountered in pairs in the canopy of the forest and forest margins, only in the lowlands, but there are few records outside the north. It has distinctive calls, which are not as harsh as those of the previous species.

CAERULEAN CUCKOO-SHRIKE
Coracina temminckii (25 cm)   Plate 14c

The Caerulean Cuckoo-shrike is an all bluish species, with black around the base of the bill, that is found in hill forests; it has a rather peculiar call.

The SLATY CUCKOO-SHRIKE *C. schistacea* is endemic to the Banggai and Sula Islands. It is slaty grey with a glossy black head and neck and blackish under-tail coverts.

PYGMY CUCKOO-SHRIKE
Coracina abbotti (19 cm)   Plate 14b

The small Pygmy Cuckoo-shrike is endemic to the mountain forests of the central regions, where it sometimes joins mixed flocks of foraging insectivorous birds. It is blue-grey above and white below, with the face and throat black in the male and grey in the female.

SULAWESI CICADABIRD
Coracina morio (21 cm)   Plate 14e

A small, dark cuckoo-shrike of lowland and hill forest, the male is slaty, black about the face and throat, but with pale margins to the flight feathers. The female is slightly paler above, and is readily identified by its rufous under-parts, barred with black.

The Sula Islands have their own endemic species, the SULA CICADABIRD *C. sula*.

SULAWESI TRILLER
Lalage leucopygialis (17 cm)   Plate 14f

Rather small, long-tailed members of the cuckoo-shrike family, trillers live in pairs in open or lightly wooded terrain, and have rather harsh chattering notes. Sulawesi has two trillers, the White-shouldered and Sulawesi Trillers. The latter, formerly treated as a subspecies of the Pied Triller *L. nigra* of the Sundaic region, is now given full species status, and is endemic.

The male Sulawesi Triller is black above and greyish-white below, with a white supercilium above the eye, a white rump, and white in the wings. In the female, the black is replaced with grey or brownish-grey, except for the black crown.

Unfortunately, the WHITE-SHOULDERED TRILLER *L. sueurii* (17 cm) is extremely similar, but the white supercilium is poorly marked, or sometimes absent. There is more white in the wing, and the under-parts are a cleaner white. In the female, the dark parts are brown.

It is believed that the Sulawesi Triller, a softer-voiced bird of well-wooded areas, is being gradually replaced from the south by the more strident White-shouldered Triller of open country, but their similarity is a constraint in unravelling their relationship.

Bulbuls

THERE are no indigenous bulbuls on mainland Sulawesi, but both the SOOTY-HEADED BULBUL *Pycnonotus aurigaster* (20 cm) and the YELLOW-VENTED BULBUL *P. goiavier* have been introduced into South Sulawesi where they occur quite widely. The endemic GOLDEN BULBUL *Hypsipetes affinis* of Maluku occurs on the Sula, Banggai, and Sangihe Islands; this is variably yellowish to greenish in its different island subspecies.

Thrushes

THE thrushes, chats, and shortwings are fine-billed insectivorous birds. Sulawesi has just seven species, including three endemics and two winter visitors. The robins and forktails of Asia do not cross Wallace's Line into Sulawesi.

GREAT SHORTWING
Heinrichia calligyna (17 cm) Plate 15a

Like all shortwings, this is a skulking bird of the undergrowth in the montane forests. It is endemic to Sulawesi and known from only the Rorokatimbu, Latimojong, Mengkoka, and Matinan Mountains. Large for a shortwing, the male is mostly deep blue and the female is a mixture of slaty-blue and reddish-brown. There is no white eyebrow which is a feature of other shortwings. However, the three subspecies on the island are very distinct, and in particular the females on Mount Matinan in the north (*H. c. simplex*) are more brown than blue-grey (see Plate 15ai). The song, described from a recording made by David Gibbs, is a hesitant song of six loud, slow, deliberate whistles, slightly wavering, repeated in the same sequence for long periods. The second note rises and falls, there is a pause after the third note, and the last two notes are more rapid.

PIED BUSH-CHAT
Saxicola caprata (13 cm)

The Pied Bush-chat is a familiar bird in Java, and has a wide range through Asia to New Guinea. It is a common bird of grassland and scrub in both the lowlands and upland valleys of Sulawesi, usually seen in pairs, perched on stones, low bushes, or telegraph wires, and catching insects on the ground or sometimes in the air. It has a habit of frequently flicking the tail. The white rump and wing patch in the black plumage of the male are distinctive. The female is brown, greyish above, with a pale rump, and only a small white spot on the bend of the wing. The song is a short though rather nondescript warble.

Pied Bush-chat

SULAWESI THRUSH
Cataponera turdoides (20 cm) Plate 15b

The forest thrushes tend to be extremely secretive, living on or close to the ground, and this is another very poorly known endemic, known only from the mountain forests of South, South-east, and Central Sulawesi. Generally olive-brown, the facial features are distinctive, although they vary somewhat between subspecies. The bill and legs are orange. It has a thrush-like chattering alarm call, and is believed to have a pleasant, melodious song.

RED-BACKED THRUSH
Zoothera erythronota (20 cm) Plate 15d

As Plate 15d shows, this endemic thrush has very distinctive plumage. It is a typical thrush, with a hopping gait on the ground and lower branches, inspecting the leaf litter on the forest floor, but it is also prone to making fast dashes. It has a liquid, thrush's song. It is believed to be quite common in the lowland forests and lower hills. There is a distinctive subspecies on Peleng Island, and probably this same subspecies has recently been discovered on Taliabu in the Sula Islands.

ISLAND THRUSH
Turdus poliocephalus (22 cm)   Plate 15c

The remarkable Island Thrush is found in the mountains of South Sulawesi, and apparently at lower altitudes around Soroako. Related to the Blackbird *T. merula* of Europe and the American Robin *T. migratorious*, it is found on islands from Taiwan, the Philippines, and Indonesia through to the Pacific. On the larger islands, it occurs only on the higher mountains, but it is found at lower elevations on the smaller islands. Nearly every island in its range, and sometimes every mountain, has a distinctive subspecies. The bird illustrated in Plate 15c is *T. p. celebensis* of the Lompobattang mountains of the south. It is a typical thrush, feeding on the ground, but shy and always ready to fly off through the understorey with a chattering alarm when disturbed. The yellow bill and legs often serve to identify a bird in flight. Sulawesi birds have more or less olive upperparts and head, lighter on the breast, and a rufous belly, white in the centre. The song has been described (from birds in Java) as beginning slowly with alternating high and low notes, speeding gradually to a 'gloriously lusty song'.

The BLUE ROCK-THRUSH *Monticola solitarius* (23 cm) may be quite a common visitor to North Sulawesi, on rocky shores and also around buildings. The male is blue with a chestnut belly.

Babblers

SULAWESI BABBLER
Trichastoma celebense (15 cm)   Plate 16a

The great family of Babblers barely crosses Wallace's Line, and Sulawesi has just this one endemic species. A dull-coloured and secretive bird of thickets, both in forest and open country in the lowlands and locally in the hills, its song is heard everywhere at dawn. This song is loud and monotonous, consisting of about three to five sliding whistles, usually delivered by pairs in antiphonal duets; the partner uttering plaintive 'peeu' calls. This so closely resembles the song of Abbott's Babbler *T. abbotti* of the Sundaic region that the two species must be very closely related.

Malia and Geomalia

ALTHOUGH these two endemic birds have similar names, they are not related, and are described together for convenience only. The Malia is often placed with the babblers, and the Geomalia with the thrushes, but their actual affinities are undetermined.

MALIA
Malia grata (22 cm) Plate 16d

This unique bird is common in the higher montane forests, where small parties move actively in the understorey or about the moss on the larger branches. Its generally olive coloration recalls a bulbul, but there is no further resemblance. Although a shy bird, it is noisy, with calls recalling those of the South-East Asian laughing-thrushes, uttered with the tail cocked over the head.

GEOMALIA
Geomalia heinrichi (18 cm) Plate 16h

This is another higher montane endemic, but it appears to be very shy and uncommon. It lives close to the ground, and has weak flight; it has a rather long, slightly graduated tail. No call note has been described.

Warblers

WARBLERS are thin-billed, usually small insect-eaters of the foliage, brown, yellow, or green in colour. Many have distinctive songs. There are eight species in Sulawesi, including two endemics.

CLAMOROUS REED-WARBLER
Acrocephalus stentoreus (19 cm)   Plate 16b

The status of the Reed-warblers in Sulawesi remains far from clear. They are abundant in the south (an estimated 26,000 birds around Lake Tempe in January), and can be found in swamp vegetation elsewhere. At present, it is not known what proportion of these are resident *A. stentoreus* or migrant ORIENTAL REED-WARBLER *A. orientalis*, as the two species can only be distinguished readily in the hand. They are large warblers, brown above and paler below, with a buff eyebrow, often seen perched on reeds, uttering harsh 'tchack' notes or singing a very guttural, rattling song of single and double notes. The Clamorous Reed-Warbler can sometimes be identified by its yellow mouth lining, and the Oriental has a longer, broader eyebrow and a heavier bill.

SULAWESI LEAF-WARBLER
Phylloscopus sarasinorum (11 cm)   Plate 16e

As their name implies, leaf-warblers are tiny warblers that feed in the canopy of trees or shrubs. This brownish-olive bird with a long, pale yellowish-white eye-stripe has been reported to be the commonest small bird in the higher forests of Lore Lindu National Park, in small noisy parties with a repertoire of sometimes discordant trills. It is endemic to Sulawesi, but a closely related species occurs in the Philippines.

The little FLYEATER *Gerygone sulphurea* (9 cm) is grey-brown above, yellow below, and has a rather rounded head and thick bill. The lores, between the bill and eye, are white. It is at once identified by its wheezy song of three to five descending notes, often uttered as if with effort. It can be found in trees anywhere, from those lining town streets to the highest forests in the mountains.

CHESTNUT-BACKED BUSH-WARBLER
Bradypterus castaneus (16 cm)   Plate 16g

A bird of the higher hills, from 1000 metres to the tops of the mountains, this is a common but skulking warbler of dense thickets at ground level. It has a variety of thin, shrill notes.

MOUNTAIN TAILORBIRD
Orthotomus cuculatus (12 cm)   Plate 16f

The only species of tailorbird in Sulawesi and the Sula Islands, this is common in forest edge habitats in the mountains. It is seen usually in pairs, or little family parties, active in the scrub growth, where its tinkling, musical song is quite unlike any other tailorbird. Sulawesi birds differ from their cousins elsewhere in having a variably rufous to chestnut crown, this colour sometimes extending down to the cheeks, there is no readily visible eye-stripe, and the yellow on the belly is confined to the flanks. The tail is long and often cocked.

GOLDEN-HEADED CISTICOLA
Cisticola exilis (10 cm)   Plate 16c

These tiny warblers are common in dry, open grasslands and scrub in the lowlands and upland valleys, often seen in weak, fluttery, rather dancing flight from the top of one grass clump to another. They are very streaky brown birds, but in the breeding season, the male acquires a golden crown, which is diagnostic. The song at this season is also readily recognizable, a wheezy note followed by a hollow 'plop'.

In wet grasslands and rice-fields, it is replaced by the ZITTING CISTICOLA *C. juncidis*. This is a much more streaky bird, which acquires no golden crown, and its 'song' consists of a repeated 'zit, zit, zit' uttered in a dipping display flight.

A much larger grassland warbler with a long and rather broad tail is the TAWNY GRASSBIRD *Megalurus timoriensis* (20 cm). Visitors to Lore Lindu National Park may encounter this in the Besoa and Napu plains, the only place it has been positively confirmed in Sulawesi.

Flycatchers

MAINLAND Sulawesi has eleven species of flycatcher, all but three of which are montane birds. No less than five are endemic, and there are other species on the offshore islands. Flycatchers form a varied and often quite colourful family of small insect-eaters. They are usually found in the lower or middle levels of the forest, quietly foraging about the branches and occasionally flying out to catch insects in mid-air.

RUFOUS-THROATED FLYCATCHER
Ficedula rufigula (11 cm) Plate 17a

This is a typical 'blue flycatcher' in which the male is slaty blue above, with a rufous throat and breast, becoming white on the belly. The female is a duller, greyish bird. Like all the blue flycatchers, it is a quiet and secretive bird of the understorey, not readily seen although quite common. It is endemic to the lowland forests of Sulawesi, up to about 1000 metres.

At higher elevations, it is replaced by the SNOWY-BROWED FLYCATCHER *F. hyperythra*, which differs in having a thick white eyebrow, and also a small black chin. This is a common bird in the mountains of Sulawesi, and elsewhere in South-East Asia.

MANGROVE BLUE FLYCATCHER
Cyornis rufigastra (13 cm) Plate 17b

This flycatcher is very similar to the Rufous-throated Flycatcher. The upper-parts are a brighter blue, and the under-parts a more orange rufous. The female is also blue and rufous, but less bright. It has a typical flycatcher's quiet but not unmusical short song phrases. The name is very misleading, being derived from the more typical habitat of coastal forests frequented by this species in the Sundaic region; in Sulawesi, it is a bird of hill forest, where it is common in the middle storey. Thus, both this bird and the Rufous-throated Flycatcher may be present in the same area.

BLUE-FRONTED BLUE FLYCATCHER (Blue-fronted Flycatcher)
Cyornis hoevelli (18 cm) Plate 17c

This is a common endemic flycatcher in the mountain forests of Central and South-east Sulawesi. This male bird is less colourful than the previous species, and indeed the generic name 'blue flycatcher' is barely justified. The female has grey or grey-brown fore-parts. It has a prolonged warbling song of clearly uttered notes that rise and fall in the scale in a complex fashion. In the mountains of North Sulawesi, it is replaced by the endemic MATINAN BLUE FLYCATCHER *C. sanfordi*, in which the male resembles the female of the previous species.

South Sulawesi also has its endemic, the LOMPOBATTANG FLYCATCHER *Ficedula bonthaina*, which is apparently confined to a zone at about 1000 metres elevation on the mountain of that name. Its continued presence has only recently been confirmed, after many decades, but it may be endangered by the forest clearance that is advancing up the mountain slopes. It is olive-brown above, paler below, with a chestnut tail, and a pale spot in front of the eye. It is related to the Cryptic Flycatcher *F. crypta* of the Philippines.

Another species which is quite common in hill forests but often overlooked is the LITTLE PIED FLYCATCHER *Ficedula westermanni*. This is a distinctive little bird, the male black above and white below, with a prominent white eye-stripe and a white bar in the wing. The female has the black parts replaced with grey and brown.

ISLAND FLYCATCHER (Island Verditer Flycatcher)
Eumyias panayensis (16 cm) Plate 17g

This rather large flycatcher of hill woodland and forest is quite common, but like all flycatchers it is readily overlooked. It is greenish-blue with a white belly, and a shining light blue throat and supercilium. Birds in the southern peninsula are rather larger and darker. It is closely related to the Verditer Flycatcher *E. thalassina* of the Sundaic region.

CITRINE FLYCATCHER (Fly-robin)
Culicicapa helianthea (12 cm) Plate 17e

This flycatcher, which is now placed in the family of Australo-Papuan Robins (Eopsaltridae), is very closely related to the Grey-headed Flycatcher *C. ceylonsis* of the Sundaic region, but lacks the grey head of that bird. A common, warbler-like bird of the middle storey in lowland and hill forests, it is olive-green above, browner on the wings and tail, and bright yellow below. It has the same song as its Sundaic cousin, four or five clear notes, each at different pitch. It is confined to Sulawesi, the Sula Islands, and the Philippines.

BLACK-NAPED MONARCH
Hypothymis azurea (14 cm) Plate 17f

The range of this familiar blue monarch flycatcher of the oriental region extends east to Sulawesi and the Sula Islands, where it is common in the lowland forests and woodlands, up to 1200 metres elevation. Compared to Sundaic birds, the Sulawesi subspecies is a slightly paler blue and lacks the black patch on the nape from which the species is named; it also has a less strongly contrasting dark breast band. The females are duller blue and lack the breast band. The typical song is a loud 'wit-wit-wit-wit' (up to seven notes), although it also has some harsher calls.

Mainland Sulawesi has no paradise-flycatchers, but Sangihe Island has the beautiful, smoky blue endemic CAERULEAN PARADISE-FLYCATCHER *Eutrichomyias rowleyi* (16 cm). Believed to be extinct since the last sighting in 1978, it was rediscovered in 1995. Critically endangered, the loss of such a beautiful bird would be very sad, and would directly result from forest clearance, demonstrating the fragility of the ecosystem of small islands.

The Talaud Islands have a race of the PHILIPPINE (or Rufous) PARADISE-FLYCATCHER *Terpsiphone cinnamomea* (18 cm), another beautiful species, known in Indonesia only from this one island group. It is chestnut all over, with a pronounced blue orbital eye-ring, and a very long tail in the male. It is quite common in all wooded localities, revealing its presence by its calls which resemble the Black-naped Monarch, described above.

RUSTY-BELLIED FANTAIL
Rhipidura teysmanni (16 cm) Plate 17d

Fantails are active birds of the understorey, which have long, broadly graduated tails, that are constantly cocked and fanned, and short, cheerful songs. The Rusty-bellied Fantail is a common endemic in the hill forest of Sulawesi and the Sula Islands. Shades of brown above, with a rufous forehead, it has a white throat separated by a black throat band from the greyish or rusty under-parts. The bird illustrated is *R. t. teysmanni* of South Sulawesi, which has more rusty under-parts. A reddish tail is its most striking feature, but with the distal half of the tail feathers dark, sometimes with pale tips. It is commonly a participant of mixed feeding flocks.

Whistlers

USUALLY placed together with the flycatchers in books that cover the Sundaic region, the whistlers are a large Australo-Papuan family characterized by their robust build, with thick heads and strong bills, and often loud clear songs. Mainland Sulawesi has three species, all endemic.

SULPHUR-BELLIED WHISTLER
Pachycephala sulfuriventer (15 cm)   Plate 17h

A rather nondescript olive to yellow bird with pale sulphur belly, this common species of the hill forests is most easily recognized by its song. Similar to that of the better known Mangrove Whistler *P. grisola* of the Sundaic forests, the song consists of a varied series of whistles that almost always finish in one, occasionally two, explosive 'whiplash' notes.

MAROON-BACKED WHISTLER
Coracornis raveni (15 cm)   Plate 17i

This whistler is a rather secretive bird of the understorey of montane forest, and is dark-coloured with blackish fore-parts and wings, a chestnut back, and olive-grey under-parts. Its call is reported to be a four-note whistle.

In the rocky moss forest of the higher mountains, the YELLOW-FLANKED WHISTLER *Hylocitrea bonensis* (15 cm) is a quiet and rather tame bird, and is rather dull coloured except for its bright yellow flanks and rufous under-tail coverts.

Flowerpeckers

THE flowerpeckers are an oriental family of tiny, colourful birds that flit constantly from bush to bush with erratic flight and short, sharp calls. They are similar to sunbirds in their habitat and behaviour, but are more stocky in build, with shorter, more stubby bills. The males have bright but mainly unglossed colours. All three species found in Sulawesi are endemic.

GREY-SIDED FLOWERPECKER
Dicaeum celebicum (8 cm)   Plate 18a

The male has all dark upper-parts, and creamy under-parts, with a large red patch on the breast and a black median line on the belly. It is common in the lowlands and hills of Sulawesi and offshore islands, including the Sula Islands, and is closely related to the Black-sided Flowerpecker *D. monticolum* of the Kalimantan mountains.

CRIMSON-CROWNED FLOWERPECKER
Dicaeum nehrkorni (9 cm)   Plate 18b

The male Crimson-crowned Flowerpecker is glossy blue-black, but with a red crown and rump. The under-parts are greyish-white with a small red breast spot and dark median line on the belly. This flowerpecker is endemic to mainland Sulawesi, and appears to be a rather uncommon bird, confined to forest in the mountains.

YELLOW-SIDED FLOWERPECKER
Dicaeum aureolimbatum (9 cm)   Plate 18c

The Yellow-sided Flowerpecker is quite a pale bird, olive above, with the throat white, belly creamy, and with deep lemon-yellow flanks. It is endemic to Sulawesi and some of its offshore islands.

Sunbirds

THE sunbirds are a family of tiny, nectarivorous birds that differ from flowerpeckers in having fine, curved bills, and glossy metallic colours in the males. Like flowerpeckers, they are very active birds around flowering trees and shrubs. Mainland Sulawesi has four species, three of which are widespread in South-East Asia, while the fourth has a Wallacean-Papuan distribution.

BLACK SUNBIRD
Nectarinia aspasia (11 cm) Plate 18d

The Black Sunbird is at once identified by its iridescent plumage that appears to be all black. In fact, it is not entirely black, and as Plate 18d shows, there are some regional variations. *N. a. grayi* in North Sulawesi, for example, has a reddish-brown mantle and breast, compared to the darker *N. a. porphyrolaema* over the rest of mainland Sulawesi. The female is olive-yellow, with grey fore-parts. It is common in gardens, open country, and scrub in the lowlands, flitting about the bushes with short, high call notes. It is found in Sulawesi, Maluku, and New Guinea.

The OLIVE-BACKED SUNBIRD *N. jugularis* and larger BROWN-THROATED SUNBIRD *Anthreptes malacensis* are both common in the same habitat. The former occurs through most of Indonesia, and the latter as far east as the Sula Islands and Flores. The Olive-backed Sunbird has unglossed olive upper-parts and a glossy blue throat, while the larger Brown-throated Sunbird has the upper-parts glossy, and the throat brown. Both species have a yellow belly. The females are olive above and yellow below. A useful distinguishing feature in both sexes of the Olive-backed Sunbird is the longer, finer bill and the white outer tail feathers.

CRIMSON SUNBIRD
Aethopyga siparaja (11 cm without tail plumes) Plate 18e

The male is a beautiful red bird with a yellow rump, and has a long purple tail in the breeding season. The female is brown, more red on the wings and back, and more yellow below. However, it has many similarities with the Crimson Myzomela (see p. 65), and should be studied closely. It prefers more wooded localities than other sunbirds, and is common in the lower hills. Sulawesi is the eastern limit of this sunbird's range.

Sangihe Island has its own endemic, the ELEGANT SUNBIRD

A. duyvenbodei. Rather similar to the previous species, it has the entire under-parts yellow, becoming orange on the sides of the breast. It appears to be restricted to the few remaining forest remnants and is another of that island's endangered birds.

White-eyes

As a family, white-eyes are readily identified by their olive plumage and prominent white ring round the eyes, and their habit of flying from tree to tree in loose parties. However, the different species are quite confusing for the amateur. Some are greener, others more yellow, while a few lack the distinguishing eye-ring from which the family is named. They are small and very active birds of the middle and upper canopy, where they keep up an incessant twittering or cheeping, not unlike the calls of baby chickens.

LEMON-BELLIED WHITE-EYE
Zosterops chloris (10 cm)   Plate 19a

This white-eye occurs mainly on small islands in the Java and Wallacean seas, but in Sulawesi it is the common species in the open and wooded country of the lowlands, including the upland valleys of Lore Lindu. It is the species most often seen in suburban gardens, but it is apparently absent from the north. Plate 19 shows that it is very yellow, including the entire under-parts.

Higher in the hills, up to the tops of the mountain, it is replaced by the MOUNTAIN WHITE-EYE *Z. montanus*, which has a greyish-white belly sharply demarcated from the yellow breast.

BLACK-FRONTED WHITE-EYE
Zosterops atrifrons (12 cm)   Plate 19b

Another common lowland species, except in the south, this white-eye has a yellow throat sharply demarcated from the white belly, and is distinguished by the black patch in front of the eye-ring. It has a distinctive shrill little warble, unusual for a white-eye. Its range extends through the Sula Islands and Seram to New Guinea.

LEMON-THROATED WHITE-EYE
Zosterops anomalus (11 cm) Plate 19c

The previous species is replaced by the endemic Lemon-throated White-eye in the southern peninsula, which is unusual in that it lacks a white eye-ring.

STREAK-HEADED DARKEYE
Lophozosterops squamiceps (13 cm) Plate 19d

It has been proposed that members of the White-eye family, other than those in the genus *Zosterops*, are called Darkeyes, because they lack the white eye-ring that is such a distinctive feature of the family (the previous species is an anomaly). The Streak-headed Darkeye has a greyish head and fore-parts, with a streaky crown, olive upper-parts, and a yellow belly. It has a song consisting of rather deliberate shrill notes. It is a common endemic throughout the mountain forests.

Honeyeaters

THE honeyeaters comprise a large Australo-Papuan family of birds that generally have fine, curved bills, for feeding on the nectar of flowering shrubs and trees, and the insects that gather there. The family includes the large and noisy friarbirds, which are so familiar in Irian Jaya, but these do not occur in Sulawesi, which lies at the western limit of the family's range. Sulawesi has just three species, two of them endemic.

CRIMSON MYZOMELA (Scarlet Honeyeater)
Myzomela dibapha (sanguinolenta) (11 cm) Plate 18f

This species is very similar to the Crimson Sunbird (see p. 63) in both appearance and behaviour, although a member of a different family. While sunbirds are generally solitary, or in pairs, the myzo-melas sometimes feed in small groups. The differences in the plumage of the males should be noted carefully in Plate 18f. The male myzomela does not have the sunbird's long tail. The females of both are russet-brown, paler below. The Crimson Myzomela is moderately common in hill forest, while the Crimson Sunbird is mostly confined to the lowlands and lower hills.

WHITE-EARED MYZA (Greater Streaked Honeyeater)
Myza sarasinorum (18 cm)　　　　　　　　　　　　　　Plate 18g

The two myzas or 'streaked honeyeaters', both of which are endemic to mainland Sulawesi, lack bright colours. They have rather elongate bodies, and nervous behaviour, scurrying about the branches of stunted, moss-draped mountain trees rather after the fashion of squirrels, uttering harsh 'kep' notes.

The White-eared Myza is very common in the higher mountains. It has a pale violet or whitish patch behind the eye. At lower elevations, and more often seen in the undergrowth, is the more secretive DARK-EARED MYZA (Lesser Streaked Honeyeater) *M. celebensis* (15 cm). This is very similar but has paler, more olive plumage, and light yellowish or pinky-orange bare skin around the eye. Its call note is similar, and it has a short song that recalls that of many small woodland birds.

Sparrows, Munias, and Finches

THESE families consist of the small, seed-eating birds, that have rounded bodies and thick bills, and include the munias that plunder the rice-fields. Sulawesi has nine species.

PALE-HEADED MUNIA
Lonchura pallida (11 cm)　　　　　　　　　　　　　　Plate 19e

This common munia in cultivation and grassland is endemic to Sulawesi and some of the Lesser Sunda Islands. It is distinguished by its white fore-parts and chestnut body, including the rump, and is similar to the White-headed Munia *L. maja* of Sumatra and Java. Although seemingly tame, these active little birds in cheeping flocks are quite wary, ever conscious of the farmer's children chasing them from field to field. It is most common in the lowlands.

BLACK-FACED MUNIA
Lonchura molucca (11 cm)   Plate 19f

The Black-faced Munia is a Wallacean endemic, ranging west to the Kangean Islands and the island of Penida off Bali. It is identified by its white rump, breast, and belly, and blackish fore-parts. At close range, the white parts, especially the belly, are seen to have fine, wavy, black barring.

The CHESTNUT MUNIA *L. malacca* (11 cm) of South-East Asia extends east to Sulawesi, Halmahera, and the Philippines. Also having a black head and fore-parts, it differs by all the rest of the body being chestnut. It is very common in Sulawesi.

SCALY-BREASTED MUNIA
Lonchura punctulata (11 cm)   Plate 19g

This munia lacks a white rump, and has heavy dark scales on all the under-parts. The upper-parts are brown, becoming somewhat rusty around the bill. It occur widely in South-East Asia, east to Sulawesi, and Nusa Tenggara.

The rather drab TREE SPARROW *Passer montanus* (15 cm) has probably arrived in Sulawesi on board ships, and is a familiar bird around towns and villages. It has now become common in many towns, and even in villages in the interior such as Wuasa and Doda in the Lore Lindu region.

The larger JAVA SPARROW *Padda oryzivora* (16 cm) has been introduced into South Sulawesi, but it is now rare. It is unmistakable with its prominent white cheeks, surrounded by black, its thick pink bill, and grey body.

SUNDA SERIN
Serinus estherae (11 cm)   Plate 19h

This interesting little bird has only recently been discovered in Sulawesi. It has isolated populations in the upper montane woodlands of North Sumatra, West Java, East Java, and Mindanao in the Philippines. It is a little yellow and grey finch, with dark wings and dark, slightly forked tail. The under-parts are whitish with black streaks. Prominent are the yellow or golden-yellow rump and upper-tail coverts, while the forehead and upper chin, and two narrow wing bars, are the same colour.

The Sunda Serin was first confirmed in Sulawesi only in 1980, above the tree-line on Mount Rantekombola in the south, and it was allocated the subspecies name *renatae*. However, subsequent sightings have been in *Agathis* forest in or around Lore Lindu National Park, at 1900 metres or above. The Lore Lindu population is unique in that the yellow parts are replaced by brilliant red-orange. This is the form illustrated in Plate 19h, and it may represent a new subspecies, as yet unnamed.

Starlings and Mynas

MEDIUM-SIZED birds, with short tails and strong, pointed bills, starlings and mynas are gregarious and generally arboreal. Their flight is swift and direct. Mainland Sulawesi has seven species, including four of Sulawesi's most distinctive endemics.

ASIAN GLOSSY STARLING
Aplonis panayensis (20 cm)   Plate 20a

The glossy starlings are sleek, glossy black birds with red eyes, that live in small parties in open, wooded country, especially in coastal coconut groves. They congregate at large evening roosts where their single, shrill calls produce a cacophony of sound. In a good light, the plumage is seen to have a green gloss. Young birds are much paler and heavily streaked.

This species is common throughout Sulawesi, but is very difficult to distinguish from the smaller SHORT-TAILED STARLING *A. minor* (18 cm), which has the fore-parts glossed purple instead of green. However, both size and the colour of a 'gloss' are often difficult to establish, and the shorter tail is probably the best aid in identification. *A. panayensis* is a widespread species in the oriental region, whereas *A. minor* appears to have a restricted relict population in the Philippines, Sulawesi, the Lesser Sundas, and, perhaps as a migrant, East Java.

WHITE-VENTED MYNA
Acridotheres javanicus (24 cm)   Plate 20c

The White-vented Myna is quite common in South Sulawesi, and now also occurs around Palu. The body, wings, and tail are grey, the head is black, and the bill, feet, and eye-ring are yellow. The tips of the tail feathers, the under-tail coverts, and a flash in the wing are white. This may be a feral population, but its source and taxonomic affinity await clarification.

FINCH-BILLED MYNA
Scissirostrum dubium (22 cm)   Plate 20b

This very gregarious and vocal bird is one of Sulawesi's most conspicuous endemics. It is unmistakable, being dark grey with prominent waxy crimson tips to many of the feathers, especially at the rear of the body. It has a thick, heavy yellow bill and yellow legs. Flocks constantly wheel noisily around the large nesting colonies, which have been established by boring out holes in dead and rotting trees in the wooded lowlands. Disaster may strike the colony when a rotting tree collapses.

WHITE-NECKED MYNA
Streptocitta albicollis (46–50 cm)   Plate 20d

With its long tail, the White-necked Myna is a rather magpie-like bird. It is another prominent endemic of wooded lowlands. Seen in twos or threes, it has a characteristic stance, with its tail often held down beneath the body. The black and white plumage is unmistakable (but compare that of the Piping Crow, see p. 72). It has a yellow-tipped bill, although *S. a. torquata* in the north and east has an all black bill. It has a variety of call notes, sometimes recalling those of a drongo, or the treepies of the oriental region.

It occurs throughout Sulawesi and its immediate offshore islands, but is replaced on the Sula Islands by another endemic, the BARE-EYED MYNA *S. albertinae*, which has much more extensive white in the plumage, the black has a greenish tone, and the sides of the head have dark bare skin.

SHORT-CRESTED (or Sulawesi Crested) MYNA
Basilornis celebensis (23 cm)　　　　　　　　　　　　　Plate 20e

This is a very distinctive bird, with its peculiarly shaped crest and the whitish patches on the sides of the neck and breast (usually stained slightly yellow or brown). It lives in small parties in forest edge habitats of the lower hills, and would not be very conspicuous except for its wide repertoire of calls.

The Short-crested Myna is confined to Sulawesi and Buton, and is replaced on the Banggai and Sula Islands by the HELMETED MYNA *B. galeatus*, which has a much larger and more prominent crest. There are just two other congeners, in Seram and the Philippines.

FIERY-BROWED MYNA
Enodes erythrophris (22 cm)　　　　　　　　　　　　　Plate 20f

This strikingly coloured endemic is confined to the hill and mountain forests of mainland Sulawesi, where it is seen in pairs or small parties. The crown often appears to be entirely brick-red, and the yellow rump and tail feathers are prominent in flight. The rather pointed tail has a creamy tip. It has fluttering flight and a twittering call.

Orioles

BLACK-NAPED ORIOLE
Oriolus chinensis (25 cm)　　　　　　　　　　　　　Plate 20g

This is the only oriole found on Sulawesi, where it is at the eastern limit of its range. It is very common, and appears to fill a wide niche from open country to lowland forest, recorded up to an altitude of 1200 metres. It is unmistakable, with its golden-yellow plumage, with a broad black bar through the face and round the nape, and its very fluty song-phrases. The subspecies in Sulawesi is smaller than those found elsewhere.

Drongos

HAIR-CRESTED DRONGO
Dicrurus hottentottus (30 cm)

Drongos are slender, generally glossy black birds which have a moderately long forked tail. The Hair-crested Drongo is very common in lowland forest, where it is readily seen and heard, with a wide range of calls. The rather heavy bill should be noted. The forked tips of the tail are sometimes up-turned. The white eyes distinguish it from the dark-eyed SPANGLED DRONGO *D. bracteatus*, which replaces it in the hill and mountain forests.

Hair-crested Drongo

Wood-swallows

IVORY-BACKED WOOD-SWALLOW
Artamus monachus (20 cm)

Wood-swallows are at once identified by their squat shape and soaring flight, on broad triangular wings. They generally occur in small, loose parties which come together to rest huddled closely on a branch. The Ivory-backed Wood-swallow is a large species confined to Sulawesi and the Sula Islands, and has a shining white back, rump, and belly, a dark head, throat, and tail, and grey wings. It is found in clearings in wooded areas, most often in the hills.

Ivory-backed Wood-swallow

The more widely distributed WHITE-BREASTED WOOD-SWALLOW *A. leucorynchus* (18 cm) is generally commoner and replaces it in the lowlands, open country and towns, and it also occurs up to 2000 metres altitude in the mountains. It has not been recorded from the Sula Islands. It is very similar, with a white belly and rump, but the back is grey, not shining white.

Crows

PIPING CROW
Corvus typicus (38 cm)

This black and white crow is one of the more distinctive endemics of Sulawesi. It is found in small, loose parties at all elevations, but most often in forest clearings in the hills. The most remarkable feature of this crow is its piping call notes, a variety of quite human and sometimes rather cheeky whistles that are remarkably loud at close range.

The SLENDER-BILLED CROW *C. enca* (46 cm) of the Sundanese region is common in the lowlands, and is all black. It has a typical crow's cawing notes. A much smaller version of this bird has been separated as an endemic species confined to the Banggai Islands, the BANGGAI CROW *C. unicolor* (39 cm), but nothing is known about its status, or whether it is still extant.

Piping Crow

Appendix

A Checklist of Resident Land Birds in the Sulawesi Faunal Region

THE following list tabulates all those species that breed or are presumed to breed in Sulawesi and its offshore islands and in the Sula Islands. Maritime birds, and those species that are presumed to be only non-breeding visitors, are excluded. Thus, many common species that are described in the text are missing from this list, such as Barn Swallow, Sacred Kingfisher, and Rainbow Bee-eater. The total number of species recorded in the region is 392 (inclusive of eight introduced species), of which 278 are listed here: 107 of these are endemic to Wallacea, of which 97 are confined to the Sulawesi faunal region, and 56 to mainland Sulawesi only (inclusive of Muna and Buton).

The following symbols are used to denote distribution:
E   Endemic to the Sulawesi faunal region
e   Endemic to Wallacea (Sulawesi, the Moluccas, and Lesser Sundas)

Main islands:
Sw   Mainland Sulawesi (includes Muna and Buton, and immediate offshore islands)
Ba   Banggai group of islands (Banggai and/or Peleng)
Sl   Sula Islands (Mangole and Taliabu, in Maluku province)

Small island groups:
ST   Sangihe and Talaud groups of islands (includes Siau)
Tj   Tanahjampea group of islands (off South Sulawesi)

Restricted ranges within mainland Sulawesi are indicated by N, C, E, SE, and S for North, Central, East, South-east, and South Sulawesi (these are physical regions, not administrative), while Sa and Ta refer to Sangihe and Talaud islands individually within the Sangihe-Talaud group.

The Sulawesi faunal region as used in this book corresponds to that defined by Andrew (1992) in the Indonesian checklist. For more details of distribution, see White and Bruce (1986). Records have also been incorporated from the following recent publications: Stones et al. (forthcoming), Dutson (1995), Bishop (1992), and Masala, Pesik, and Indrawan (forthcoming).

APPENDIX

English Name	Scientific Name	E/e	Sw	Ba	Sl	ST	Tj
Black-throated Little Grebe	Tachybaptus novaehollandiae					X	
Red-throated Little Grebe	Tachybaptus ruficollis		X		X		
Little Black Cormorant	Phalacrocorax sulcirostris		X		X		
Little Pied Cormorant	Phalacrocorax melanoleucos		X	X	X		X
Oriental Darter	Anhinga melanogaster		X				
Great-billed Heron	Ardea sumatrana		X	X	X	X	X
Purple Heron	Ardea purpurea		X	X			
Pied Heron	Egretta picata		X				
Intermediate Egret	Egretta intermedia		X		X	Sa	
Little Egret	Egretta garzetta		X		X	Sa	
Reef Egret	Egretta sacra		X	X	X	X	X
Cattle Egret	Bubulcus ibis		X	X	X	X	X
Javan Pond-heron	Ardeola speciosa		X				X
Striated Heron	Butorides striatus		X	X	X	X	X
Black-crowned Night-heron	Nycticorax nycticorax		X				
Rufous Night-heron	Nycticorax caledonicus		X	X		X	X
Cinnamon Bittern	Ixobrychus cinnamomeus		X		X		
Black Bittern	Ixobrychus flavicollis		X		X	X	
Milky Stork	Mycteria cinerea		X				
Woolly-necked Stork	Ciconia episcopus		X	X			
Glossy Ibis	Plegadis falcinellus		X				
Osprey	Pandion haliaetus		X	X	X	X	X
Jerdon's Baza	Aviceda jerdoni		X	X	X		
Pacific Baza	Aviceda subcristata						X
Barred Honey-buzzard	Pernis celebensis		X	X			
Bat Hawk	Macheiramphus alcinus		C				
Black-winged Kite	Elanus caeruleus		X				X
Black Kite	Milvus migrans		X				
Brahminy Kite	Haliastur indus		X	X	X	X	X
White-bellied Sea-eagle	Haliaeetus leucogaster		X	X	X	X	X
Lesser Fish-eagle	Ichthyophaga humilis		X	X	X		
Grey-headed Fish-eagle	Ichthyophaga ichthyaetus		X				
Sulawesi Serpent-eagle	Spilornis rufipectus	E	X	X	X	X	X
Spotted Harrier	Circus assimilis		X		X		
Sulawesi Goshawk	Accipiter griseiceps	E	X				
Spot-tailed Goshawk	Accipiter trinotatus	E	X				
Brown Goshawk	Accipiter fasciatus						X
Small Sparrow-hawk	Accipiter nanus	E	X				

APPENDIX

English Name	Scientific Name	E/e	Sw	Ba	Sl	ST	Tj
Vinous-breasted Sparrow-hawk	*Accipiter rhodogaster*	E	X	X	X		
Rufous-winged Buzzard	*Butastur liventer*		X				
Black Eagle	*Ictinaetus malayensis*		X	X			
Rufous-bellied Eagle	*Hieraaetus kienerii*		X		X		
Sulawesi Hawk-eagle	*Spizaetus lanceolatus*	E	X	X	X		
Spotted Kestrel	*Falco moluccensis*		X	X	X	X	X
Oriental Hobby	*Falco severus*		X	X	X		
Peregrine Falcon	*Falco peregrinus*		X		X		X
Spotted Whistling-duck	*Dendrocygna guttata*		N,E		X	X	
Wandering Whistling-duck	*Dendrocygna arcuata*		X		X	Ta	
Sunda Teal	*Anas gibberifrons*		X	X	X		X
Pacific Black Duck	*Anas superciliosa*		X				
Philippine Scrubfowl	*Megapodius cumingii*		X		X		
Sula Scrubfowl	*Megapodius bernsteinii*	E		X	X		
Orange-footed Scrubfowl	*Megapodius reinwardt*						X
Maleo	*Macrocephalon maleo*	E	X				
Blue-breasted Quail	*Coturnix chinensis*		X		X		
Red Junglefowl	*Gallus gallus*		X		?	Sa	
Red-backed Button-quail	*Turnix maculosa*		X	X			
Barred Button-quail	*Turnix suscitator*		X				
Buff-banded Rail	*Gallirallus philippensis*		X		X		X
Barred Rail	*Gallirallus torquatus*		X	X	X		
Slaty-legged Crake	*Rallina eurizonoides*		X	X	X		
Snoring Rail	*Aramidopsis plateni*	E	X				
Blue-faced Rail	*Gymnocrex rosenbergii*	E	X	X			
Baillon's Crake	*Porzana pusilla*		N				
Ruddy-breasted Crake	*Porzana fusca*		X				
White-browed Crake	*Poliolimnas cinerea*		X		X		X
Common Bush-hen	*Amaurornis olivacea*				X	Sa	
Isabelline Bush-hen	*Amaurornis isabellina*	E	X				
White-breasted Waterhen	*Amaurornis phoenicurus*		X	X	X	X	X
Dusky Moorhen	*Gallinula tenebrosa*		X		X		
Common Moorhen	*Gallinula chloropus*		X			Sa	
Purple Swamphen	*Porphyrio porphyrio*		X			Ta	
Comb-crested Jacana	*Irediparra gallinacea*		X				
Malaysian Plover	*Charadrius peronii*		X		X		X
Sulawesi Woodcock	*Scolopax celebensis*	E	X				
White-headed Stilt	*Himantopus leucocephalus*		X				X

(*continued*)

APPENDIX

English Name	Scientific Name	E/e	Sw	Ba	Sl	ST	Tj
Beach Thick-knee	*Esacus magnirostris*		X		X		X
Grey-cheeked Green Pigeon	*Treron griseicauda*		X	X	X	Sa	X
Pink-necked Green Pigeon	*Treron vernans*		X	X		Ta	X
Red-eared Fruit-dove	*Ptilinopus fischeri*	E	X				
Maroon-chinned Fruit-dove	*Ptilinopus subgularis*	E	X	X	X		
Superb Fruit-dove	*Ptilinopus superbus*		X				
Black-naped Fruit-dove	*Ptilinopus melanospila*		X	X	X	X	X
White-bellied Imperial Pigeon	*Ducula forsteni*	E	X		X		
Grey-headed Imperial Pigeon	*Ducula radiata*	E	X				
Green Imperial Pigeon	*Ducula aenea*		X	X	X	Ta	X
Blue-headed Imperial Pigeon	*Ducula concinna*					X	X
Pink-headed Imperial Pigeon	*Ducula rosacea*						X
Grey Imperial Pigeon	*Ducula pickeringii*					Ta	
Pied Imperial Pigeon	*Ducula bicolor*		X			X	X
White Imperial Pigeon	*Ducula luctuosa*	E	X	X	X		
Sombre Pigeon	*Cryptophaps poecilorrhoa*	E	X				
Rock Pigeon	*Columba livia*		S				
Metallic Pigeon	*Columba vitiensis*		N	X	X		
Red Collared Dove	*Streptopelia tranquebarica*		C				
Spotted Dove	*Streptopelia chinensis*		X	X	X	X	X
Zebra Dove	*Geopelia striata*		X				X
Sulawesi Black Pigeon	*Turacoena manadensis*	E	X	X	X		
Slender-billed Cuckoo-dove	*Macropygia amboinensis*		X	X	X	X	X
Dusky Cuckoo-dove	*Macropygia magna*	e					X
Emerald Dove	*Chalcophaps indica*		X	X	X	X	X
Stephan's Dove	*Chalcophaps stephani*		X		X		
Sulawesi Ground-dove	*Gallicolumba tristigmata*	E	X				
Nicobar Pigeon	*Caloenas nicobarica*		X	X	X	Sa	
Red-and-Blue Lory	*Eos histrio*	E				X	
Ornate Lorikeet	*Trichoglossus ornatus*	E	X	X			
Rainbow Lorikeet	*Trichoglossus haematodus*						X
Yellow-and-Green Lorikeet	*Trichoglossus flavoviridis*	E	X		X		
Yellow-crested Cockatoo	*Cacatua sulphurea*	e	X				X
Red-spotted Racquet-tail	*Prioniturus flavicans*	E	N,C	?			
Golden-mantled Racquet-tail	*Prioniturus platurus*	E	X	X	X	X	
Blue-naped Parrot	*Tanygnathus lucionensis*					Ta	
Azure-rumped Parrot	*Tanygnathus sumatranus*		X	X	X	X	
Great-billed Parrot	*Tanygnathus megalorynchos*					X	X

APPENDIX

English Name	Scientific Name	E/e	Sw	Ba	Sl	ST	Tj
Moluccan King Parrot	*Alisterus amboinensis*			X	X		
Sulawesi Hanging-parrot	*Loriculus stigmatus*	E	X				
Moluccan Hanging-parrot	*Loriculus amabilis*	e		X	X		
Sangihe Hanging-parrot	*Loriculus catamene*	E				Sa	
Red-billed Hanging-parrot	*Loriculus exilis*	E	X				
Sulawesi Hawk-cuckoo	*Cuculus crassirostris*	E	X				
Plaintive Cuckoo	*Cacomantis merulinus*		X				
Rusty-breasted Cuckoo	*Cacomantis sepulcralis*		X	X	X		
Gould's Bronze Cuckoo	*Chrysococcyx russatus*		X				X
Drongo Cuckoo	*Surniculus lugubris*		X				
Asian Koel	*Eudynamys scolopacea*					X	
Black-billed Koel	*Eudynamys melanorhyncha*	E	X	X	X		
Channel-billed Cuckoo	*Scythrops novaehollandiae*		X		X	Ta	
Yellow-billed Malkoha	*Rhamphococcyx calyorhynchus*	E	X				
Lesser Coucal	*Centropus bengalensis*		X	X	X	X	X
Bay Coucal	*Centropus celebensis*	E	X				
Barn Owl	*Tyto alba*						X
Sulawesi Owl	*Tyto rosenbergii*	E	X	X		Sa	
Minahassa Masked Owl	*Tyto inexspectata*	E	N,C				
Taliabu Masked Owl	*Tyto nigrobrunnea*	E			X		
Eastern Grass-owl	*Tyto longimembris*		S,E				
Sulawesi Scops Owl	*Otus manadensis*	E	X				
Moluccan Scops Owl	*Otus magicus*			X	X	Sa	
Ochre-bellied Boobook	*Ninox ochracea*	E	X				
Speckled Boobook	*Ninox punctulata*	E	X				
Satanic Nightjar	*Eurostopodus diabolicus*	E	N				
Great-eared Nightjar	*Eurostopodus macrotis*		X	X	X	X	
Sulawesi Nightjar	*Caprimulgus celebensis*	E	N,C	X			
Large-tailed Nightjar	*Caprimulgus macrurus*						X
Savanna Nightjar	*Caprimulgus affinis*		X				
Edible-nest Swiftlet	*Aerodramus fuciphagus*						X
Uniform Swiftlet	*Aerodramus vanikorensis*		X				
Moluccan Swiftlet	*Aerodramus infuscatus*	e	X		X		
Glossy Swiftlet	*Collocalia esculenta*		X	X	X	X	X
Purple Needletail	*Hirundapus celebensis*		N,C				
Little Swift	*Apus affinis*		X				
Asian Palm-swift	*Cypsiurus balasiensis*		S,SE				
Grey-rumped Tree-swift	*Hemiprocne longipennis*		X	X	X		X

(continued)

APPENDIX

English Name	Scientific Name	E/e	Sw	Ba	Sl	ST	Tj
Common Kingfisher	*Alcedo atthis*		X	X	X	X	X
Blue-eared Kingfisher	*Alcedo meninting*		X	X	X		
Variable Dwarf Kingfisher	*Ceyx lepidus*				X		
Sulawesi Dwarf Kingfisher	*Ceyx fallax*	E	X			Sa	
Black-billed Kingfisher	*Pelargopsis melanorhyncha*	E	X	X	X		
Lilac-cheeked Kingfisher	*Cittura cyanotis*	E	X			Sa	
Ruddy Kingfisher	*Halcyon coromanda*		X	X	X	X	
Collared Kingfisher	*Halcyon chloris*		X	X	X	X	X
Talaud Kingfisher	*Halcyon enigma*	E				Ta	
Green-backed Kingfisher	*Actenoides monachus*	E	X				
Scaly-breasted Kingfisher	*Actenoides princeps*	E	X				
Blue-tailed Bee-eater	*Merops philippinus*		X				
Purple-bearded Bee-eater	*Meropogon forsteni*	E	X				
Purple-winged Roller	*Coracias temminckii*	E	X				
Common Dollarbird	*Eurystomus orientalis*		X	X	X	X	X
Sulawesi Hornbill	*Penelopides exarhatus*	E	X				
Knobbed Hornbill	*Rhyticeros cassidix*	E	X				
Ashy Woodpecker	*Mulleripicus fulvus*	E	X				
Sulawesi Woodpecker	*Dendrocopos temminckii*	E	X				
Red-bellied Pitta	*Pitta erythrogaster*		X			X	
Sula Pitta	*Pitta dohertyi*	E		X	X		
Hooded Pitta	*Pitta sordida*		N			Sa	
Elegant Pitta	*Pitta elegans*	e		X	X		X
Pacific Swallow	*Hirundo tahitica*		X	X	X	X	X
Richard's Pipit	*Anthus novaeseelandiae*		X		X		
Slaty Cuckoo-shrike	*Coracina schistacea*	E		X	X		
Caerulean Cuckoo-shrike	*Coracina temminckii*	E	X				
Pied Cuckoo-shrike	*Coracina bicolor*	E	X			Sa	
White-rumped Cuckoo-shrike	*Coracina leucopygia*	E	X				
Pygmy Cuckoo-shrike	*Coracina abbotti*	E	X				
Common Cicadabird	*Coracina tenuirostris*			X			X
Sula Cicadabird	*Coracina sula*	E			X		
Sulawesi Cicadabird	*Coracina morio*	E	X			X	
Sulawesi Triller	*Lalage leucopygialis*	E	X	X	X		
White-shouldered Triller	*Lalage sueurii*		X				X
Sooty-headed Bulbul	*Pycnonotus aurigaster*		S				
Yellow-vented Bulbul	*Pycnonotus goiavier*		S				
Golden Bulbul	*Hypsipetes affinis*	e		X	X	Sa	

APPENDIX

English Name	Scientific Name	E/e	Sw	Ba	Sl	ST	Tj
Great Shortwing	*Heinrichia calligyna*	E	X				
Pied Bush-chat	*Saxicola caprata*		X				X
Geomalia	*Geomalia heinrichi*	E	X				
Red-backed Thrush	*Zoothera erythronota*	E	X	X	X		
Sulawesi Thrush	*Cataponera turdoides*	E	X				
Island Thrush	*Turdus poliocephalus*		X				
Sulawesi Babbler	*Trichastoma celebense*	E	X				
Malia	*Malia grata*	E	X				
Chestnut-backed Bush-warbler	*Bradypterus castaneus*	e	X				
Tawny Grassbird	*Megalurus timoriensis*		C				
Clamorous Reed-warbler	*Acrocephalus stentoreus*		S				
Zitting Cisticola	*Cisticola juncidis*		X	X			
Golden-headed Cisticola	*Cisticola exilis*		X	X			
Mountain Tailorbird	*Orthotomus cuculatus*		X		X		
Sulawesi Leaf-warbler	*Phylloscopus sarasinorum*	E	X				
Island Leaf-warbler	*Phylloscopus poliocephalus*				X		
Henna-tailed Rhinomyias	*Rhinomyias colonus*	E	X	X	X		
Island Flycatcher	*Eumyias panayensis*		X		X		
Snowy-browed Flycatcher	*Ficedula hyperythra*		X		X		
Rufous-throated Flycatcher	*Ficedula rufigula*	E	X				
Lompobattang Flycatcher	*Ficedula bonthaina*	E	S				
Little Pied Flycatcher	*Ficedula westermanni*		X		X		
Blue-fronted Blue Flycatcher	*Cyornis hoevelli*	E	C,SE				
Matinan Blue Flycatcher	*Cyornis sanfordi*	E	N				
Mangrove Blue Flycatcher	*Cyornis rufigastra*		X				X
Citrine Flycatcher	*Culicicapa helianthea*		X	X	X		X
Flyeater	*Gerygone sulphurea*		X	X			X
Rufous-sided Gerygone	*Gerygone dorsalis*	e					X
Black-naped Monarch	*Hypothymis azurea*		X	X	X		
Caerulean Paradise-flycatcher	*Eutrichomyias rowleyi*	E				Sa	
Philippine Paradise-flycatcher	*Terpsiphone cinnamomea*					Ta	
Island Monarch	*Monarcha cinerascens*			X	X	X	X
White-tipped Monarch	*Monarcha everetti*	E	.				X
Broad-billed Flycatcher	*Myiagra ruficollis*						X
Rusty-bellied Fantail	*Rhipidura teysmanni*	E	X		X		
Rufous Fantail	*Rhipidura rufifrons*						X
Yellow-flanked Whistler	*Hylocitrea bonensis*	E	X				
Maroon-backed Whistler	*Coracornis raveni*	E	X				

(continued)

APPENDIX

English Name	Scientific Name	E/e	Sw	Ba	Sl	ST	Tj
Sulphur-bellied Whistler	*Pachycephala sulfuriventer*	E	X				
Common Golden Whistler	*Pachycephala pectoralis*			X	X		X
Drab Whistler	*Pachycephala griseonota*	e			X		
Yellow-sided Flowerpecker	*Dicaeum aureolimbatum*	E	X			Sa	
Crimson-crowned Flowerpecker	*Dicaeum nehrkorni*	E	X				
Red-chested Flowerpecker	*Dicaeum maugei*	e					X
Grey-sided Flowerpecker	*Dicaeum celebicum*	E	X	X	X	X	
Brown-throated Sunbird	*Anthreptes malacensis*		X	X	X	Sa	
Black Sunbird	*Nectarinia aspasia*		X	X	X	X	
Olive-backed Sunbird	*Nectarinia jugularis*		X	X	X		X
Elegant Sunbird	*Aethopyga duyvenbodei*	E				Sa	
Crimson Sunbird	*Aethopyga siparaja*		X				
Everett's White-eye	*Zosterops everetti*					Ta	
Mountain White-eye	*Zosterops montanus*		X		X		
Lemon-bellied White-eye	*Zosterops chloris*		X				X
Pale-bellied White-eye	*Zosterops consobrinorum*	E	SE				
Lemon-throated White-eye	*Zosterops anomalus*	E	S				
Black-fronted White-eye	*Zosterops atrifrons*		X	X	X	Sa	
Streak-headed Darkeye	*Lophozosterops squamiceps*	E	X				
Crimson Myzomela	*Myzomela dibapha*		X		X		X
Dark-eared Myza	*Myza celebensis*	E	X				
White-eared Myza	*Myza sarasinorum*	E	X				
Sunda Serin	*Serinus estherae*		C,S				
Tawny-breasted Parrot-finch	*Erythrura hyperythra*		X				
Blue-faced Parrot-finch	*Erythrura trichroa*		C				
Black-faced Munia	*Lonchura molucca*		X	X	X	X	X
Scaly-breasted Munia	*Lonchura punctulata*		X	X			
Chestnut Munia	*Lonchura malacca*		X				
Pale-headed Munia	*Lonchura pallida*	e	C,S				X
Java Sparrow	*Padda oryzivora*		X				
Tree Sparrow	*Passer montanus*		X	X	X		
Moluccan Starling	*Aplonis mysolensis*		E	X	X		
Short-tailed Starling	*Aplonis minor*		C,S				X
Metallic Starling	*Aplonis metallica*				X		
Asian Glossy Starling	*Aplonis panayensis*		N,C			X	
White-vented Myna	*Acridotheres javanicus*		S				
Short-crested Myna	*Basilornis celebensis*	E	X				
Helmeted Myna	*Basilornis galeatus*	E		X	X		

APPENDIX

English Name	Scientific Name	E/e	Sw	Ba	Sl	ST	Tj
White-necked Myna	*Streptocitta albicollis*	E	X				
Bare-eyed Myna	*Streptocitta albertinae*	E			X		
Fiery-browed Myna	*Enodes erythrophris*	E	X				
Finch-billed Myna	*Scissirostrum dubium*	E	X	X			
Black-naped Oriole	*Oriolus chinensis*		X	X	X	X	X
Sulawesi Drongo	*Dicrurus montanus*	E	X				
Hair-crested Drongo	*Dicrurus hottentottus*		X	X	X	X	X
White-breasted Wood-swallow	*Artamus leucorynchus*		X	X			X
Ivory-backed Wood-swallow	*Artamus monachus*	E	X	X	X		
Slender-billed Crow	*Corvus enca*		X	?	X		X
Banggai Crow	*Corvus unicolor*	E		X			
Piping Crow	*Corvus typicus*	E	X				

Among maritime birds, the following may breed or formerly have bred on some offshore islands: Great Frigatebird *Fregata minor*, Black-naped Tern *Sterna sumatrana*, Bridled Tern *Sterna anaethetus*, Lesser Crested Tern *Sterna bengalensis*, and Brown Noddy *Anous stolidus*. The Little Tern *Sterna albifrons* has bred recently on mainland Sulawesi.

Further Reading

THE serious ornithologist will need access to a more comprehensive guide than can be provided here, and a handbook to the Birds of Wallacea (Coates and Bishop) is in preparation for early publication. *Kukila* or the Bulletin of the Indonesian Ornithological Society is an international medium for the publication of papers on all aspects of Indonesian birds; it is published every year, and several papers pertaining to the Sulawesi faunal region are contained in back issues. It can be obtained from P.O. Box 310, Bogor 16003, Indonesia. Finally, reference is made in the text to the following publications:

Andrew, P. (1992), *The Birds of Indonesia: A Checklist (Peter's Sequence)*, *Kukila* Checklist No. 1, Jakarta, Indonesian Ornithological Society.

Bishop, K. D. (1992), 'New and Interesting Records of Birds in Wallacea', *Kukila*, 6(1).

Coates, B. J. and Bishop, K. D. (forthcoming), *A Guide to the Birds of Wallacea: Sulawesi, Moluccas and Lesser Sundas*, Dove Publications.

Dutson, G. (1995), 'The Birds of Salayar and the Flores Sea Islands', *Kukila*, 7(2).

Masala, I., Pesik, L., and M. Indrawan (forthcoming), 'Interesting Observations from the Banggai Islands in 1991', *Kukila*.

Stones, A. T., Lucking, R. S., Davidson, P. J., and Wahyu Rahajaningtrah (forthcoming), 'Checklist of the Birds of the Sula Islands (1991) with Particular Reference to Taliabu Island', *Kukila*.

White, C. M. N. and Bruce, M. D. (1986), *The Birds of Wallacea (Sulawesi, the Moluccas and Lesser Sunda Islands, Indonesia). An Annotated Check-list*, London, British Ornithologists Union (Check-list 7).

Index to Genera, Systematic Section

References in bold refer to Colour Plate numbers.

ACCIPITER, 12, **2**
Acridotheres, 69, **20**
Acrocephalus, 56, **16**
Actenoides, 43, **11**
Actitis, 20
Aerodramus, 38
Aethopyga, 63, **18**
Alcedo, 41, **11**
Amaurornis, 18, **5**
Anas, 13, **3**
Anhinga, 2
Anthreptes, 63
Anthus, 49
Aplonis, 68, **20**
Apus, 39
Aramidopsis, 17, **4**
Ardea, 4
Ardeola, 4
Artamus, 71

BABBLER, 54, **16**
Basilornis, 70, **20**
Bee-eaters, 44, **12**
Bitterns, 6, **1**
Boobooks, 36, **10**
Booby, 3
Bradypterus, 56, **16**
Bubulcus, 4
Bulbuls, 52
Bush-chat, 53
Bush-hen, 18, **5**
Bush-warbler, 56, **16**
Butastur, 11
Butorides, 5, **1**
Button-quails, 16
Buzzards, 11

CACATUA, 29, **8**
Cacomantis, 32, **9**
Caloenas, 28, **7**
Caprimulgus, 37, **10**
Casmerodius, 5
Cataponera, 53, **15**
Centropus, 34, **9**
Ceyx, 42, **11**
Chalcophaps, 28, **7**
Charadrius, 20
Chat, 53
Chlidonias, 23
Chrysococcyx, 33, **9**
Cicadabirds, 51, **14**
Ciconia, 7
Circus, 12, **2**
Cisticola, 57, **16**
Cisticola, 57, **16**
Cittura, 43, **11**
Cockatoo, 29, **8**
Collocalia, 38
Coracias, 45, **12**
Coracina, 50, **14**
Coracornis, 61, **17**
Cormorants, 2
Corvus, 72
Coturnix, 16
Coucals, 34, **9**
Crakes, 17, **4**
Crows, 72
Cryptophaps, 26, **7**
Cuckoos, 32, **9**
Cuckoo-doves, 27, **7**
Cuckoo-shrikes, 50, **14**
Cuculus, 32, **9**
Culicicapa, 59, **17**

INDEX TO GENERA, SYSTEMATIC SECTION

Curlews, 20
Cyornis, 58, **17**
Cypsiurus, 39

DARKEYE, 65, **19**
Darter, 2
Dendrocopos, 47, **13**
Dendrocygna, 13, **3**
Dicaeum, 62, **18**
Dicrurus, 71
Dollarbird, 46, **12**
Doves, 23, **6**, **7**
Drongos, 71
Ducks, 13, **3**
Ducula, 25, **6**

EAGLES, 11, **2**
Egrets, 4
Egretta, 5, **1**
Elanus, 10
Enodes, 70, **20**
Eos, 30
Esacus, 21
Eudynamys, 33, **9**
Eumyias, 59, **17**
Eurostopodus, 37, **10**
Eurystomus, 46, **12**
Eutrichomyias, 60

FALCO, 12, **2**
Fantail, 60, **17**
Ficedula, 58, **17**
Finches, 66
Flowerpeckers, 62, **18**
Flycatchers, 58, **17**
Flyeater, 56
Fregata, 1
Frigatebirds, 1
Fruit-doves, 24, **6**

GALLICOLUMBA, 28, **7**
Gallinula, 18, **5**
Gallirallus, 16, **4**

Gallus, 16
Garganey, 14, **3**
Geomalia, 55, **16**
Geomalia, 55, **16**
Gerygone, 56
Godwits, 20
Goshawks, 12, **2**
Grassbird, 57
Grass-owl, 36
Grebe, 1, **3**
Green Pigeons, 23, **6**
Ground-dove, 28, **7**
Gymnocrex, 17, **4**

HALCYON, 40, **11**
Haliaeetus, 9
Haliastur, 10
Hanging-parrots, 30, **8**
Harrier, 12, **2**
Hawk-cuckoo, 32
Hawk-eagle, 11, **2**
Hawk-owls, 36, **10**
Heinrichia, 52, **15**
Hemiprocne, 39
Herons, 4
Himantopus, 21
Hirundapus, 39
Hirundo, 48
Honey-buzzard, 11, **2**
Honeyeaters, 65, **18**
Hornbills, 46, **13**
Hylocitrea, 61, **17**
Hypothymis, 60, **17**
Hypsipetes, 52

IBIS, 7
Ictinaetus, 11
Imperial Pigeons, 25, **6**
Irediparra, 19, **5**
Ixobrychus, 6, **1**

JACANA, 19, **5**
Junglefowl, 16

INDEX TO GENERA, SYSTEMATIC SECTION

KESTREL, 12, **2**
Kingfishers, 40, **11**
Kites, 9
Koel, 33, **9**

LALAGE, 51, **14**
Leaf-warbler, 56, **16**
Lonchura, 66, **19**
Lophozosterops, 65, **19**
Loriculus, 30, **8**
Lorikeets, 29, **8**
Lory, 30

MACROCEPHALON, 15, **Frontispiece**
Macropygia, 27, **7**
Maleo, 15, **Frontispiece**
Malia, 55, **16**
Malia, 55, **16**
Malkoha, 34, **9**
Megalurus, 57
Megapodius, 16, **4**
Meropogon, 45, **12**
Merops, 44, **12**
Milvus, 9
Monarch, 60, **17**
Monticola, 54
Moorhens, 18, **5**
Motacilla, 49
Mulleripicus, 47, **13**
Munias, 66, **19**
Mycteria, 7
Mynas, 69, **20**
Myza, 66, **18**
Myza, 66, **18**
Myzomela, 65, **18**
Myzomela, 65, **18**

NECTARINIA, 63, **18**
Needletails, 39
Night-herons, 5, **1**
Nightjars, 37, **10**
Ninox, 36, **10**
Nycticorax, 5, **1**

ORIOLE, 70, **20**
Oriolus, 70, **20**
Orthotomus, 57, **16**
Osprey, 8, **2**
Otus, 36, **10**
Owls, 35, **10**

PACHYCEPHALA, 61, **17**
Padda, 67
Palm-swift, 39
Pandion, 8, **2**
Paradise-flycatchers, 60
Parrots, 29, **8**
Passer, 67
Pelargopsis, 42, **11**
Penelopides, 46, **13**
Pernis, 11, **2**
Phalacrocorax, 2
Phalarope, 20
Phalaropus, 20
Phylloscopus, 56, **16**
Pigeons, 23, **6**, **7**
Pipit, 49
Pittas, 47, **13**
Pitta, 47, **13**
Plegadis, 7
Plovers, 20
Pluvialis, 20
Poliolimnas, 17, **4**
Pond-heron, 4
Porphyrio, 19, **5**
Prioniturus, 31, **8**
Ptilinopus, 24, **6**
Pycnonotus, 52

QUAILS, 16

RACQUET-TAILS, 31, **8**
Rails, 16, **4**
Redshank, 20
Reed-warblers, 56, **16**
Rhamphococcyx, 34, **9**
Rhipidura, 60, **17**

INDEX TO GENERA, SYSTEMATIC SECTION

Rhyticeros, 46, **13**
Rock-thrush, 54
Rollers, 45, **12**

SANDPIPERS, 20
Saxicola, 53
Scissirostrum, 69, **20**
Scolopax, 20, **5**
Scops Owls, 36, **10**
Scrubfowl, 16, **4**
Scythrops, 34, **9**
Sea-eagle, 9
Serin, 67, **19**
Serinus, 67, **19**
Serpent-eagle, 11, **2**
Shortwing, 52, **15**
Sparrows, 67
Sparrow-hawks, 12, **2**
Spilornis, 11, **2**
Spizaetus, 11, **2**
Starlings, 68, **20**
Sterna, 22
Stilt, 21
Stint, 19
Storks, 7
Streptocitta, 69, **20**
Streptopelia, 27, **7**
Sula, 3
Sunbirds, 63, **18**
Surniculus, 33, **9**
Swallows, 48
Swamphen, 19, **5**
Swiftlets, 38
Swifts, 38–9

TACHYBAPTUS, 1, **3**
Tailorbird, 57, **16**
Tanygnathus, 31, **8**
Teal, 13, **3**
Terns, 22
Terpsiphone, 60
Thick-knee, 21
Thrushes, 53, **15**
Tree-swift, 39
Treron, 23, **6**
Trichastoma, 54, **16**
Trichoglossus, 29, **8**
Trillers, 51, **14**
Tringa, 20
Turacoena, 27, **7**
Turdus, 54, **15**
Turnix, 16
Tyto, 35, **10**

WADERS, 19
Wagtails, 49
Warblers, 56, **16**
Waterhen, 18, **5**
Whimbrel, 20
Whistlers, 61, **17**
Whistling-duck, 13, **3**
White-eyes, 64, **19**
Woodcock, 20, **5**
Woodpeckers, 47, **13**
Wood-swallows, 71–2

ZOOTHERA, 53, **15**
Zosterops, 64, **19**